Second Edition

WATER QUALITY INDICATORS GUIDE

Surface Waters

DAKOTA COUNTY LIBRARY
1340 WESCOTT ROAD
EAGAN, MINNESOTA 55123-1099

Charles R. Terrell
National Environmental Coordinator
Biological Conservation Sciences Division
Natural Resources Conservation Service
Washington, D.C.

Dr. Patricia Bytnar Perfetti, Head
Department of Geoscience and
Environmental Studies
Physics and Astronomy
University of Tennessee—Chattanooga
Chattanooga, Tennessee

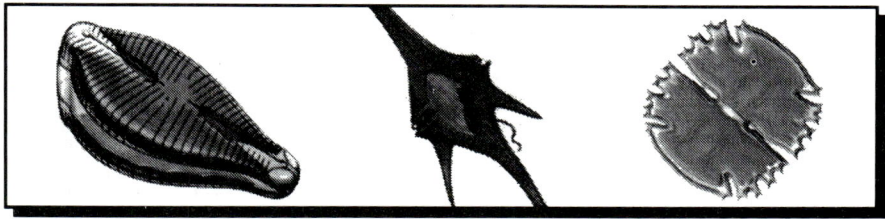

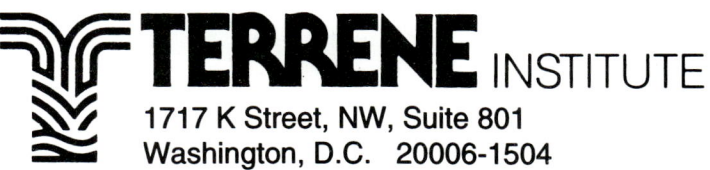

TERRENE INSTITUTE
1717 K Street, NW, Suite 801
Washington, D.C. 20006-1504

KENDALL/HUNT PUBLISHING COMPANY
4050 Westmark Drive P.O. Box 1840 Dubuque, Iowa 52004-1840

Issued September, 1989
Second Edition January 1996.

For additional copies of this publication, contact the Terrene Institute, 1717 K Street, NW, Suite 801, Washington, DC 20006-1504. Telephone 202 833-8317, fax 202 296-4071. Suggested retail price is $26.95 plus $4 shipping and handling. This suggested retail price is subject to change.

Points of view expressed in this publication do not necessarily reflect the views or policies of the Terrene Institute, Kendall/Hunt Publishing Company or the USDA Natural Resources Conservation Service. Mention of trade names and commercial products does not constitute endorsement of their use.

Copyright © 1989, 1996 by the Terrene Institute

ISBN 0-7872-1787-6

All rights reserved. No part of this publication may be reproduced, stored in a retrieval system, or transmitted, in any form or by any means, electronic, mechanical, photocopying, recording, or otherwise, without the prior written permission of the copyright owner.

Printed in the United States of America
10 9 8 7 6 5 4 3 2 1

FOREWORD TO THE SECOND EDITION

Within the past year the Soil Conservation Service (SCS) has been renamed as the Natural Resources Conservation Service (NRCS). However, this has not been merely a change in name; our mission has become permanently changed to include all natural resources, not just soil.

While SCS has been active for years in areas other than soils, such as water issues, the new charge officially recognizes NRCS' roles in soils, water, animals, plants, air and cultural resources. Additionally, new emphases are being placed within and without NRCS to integrate resources into holistic, ecological approaches, such as those employed in the *Water Quality Indicators Guide: Surface Waters*.

Perhaps the *Water Quality Indicators Guide* was ahead of its time, but the book was written with an ecological, integrated mode in mind. The authors realized that water wasn't of poor quality simply because it was located at a specific place at a specific time. They recognized that waters were affected by activities that occurred *away* from the watercourse or water body. Those distant activities and conditions became the causes of the effects that were exhibited in the water.

That concept, although simple in explanation, had eluded many authors and investigators for years. Now, we understand that when assessing water quality, we must examine the entire watershed for causes of water degradation. To improve water quality it is necessary to find those causes of degradation, and to address and correct them.

The *Water Quality Indicators Guide* has withstood the test of time. It has shown that an ecological approach to assessing water quality is a vital tool in the overall tool box of assessment technologies. This second edition updates the first edition and again makes this significant document available to individuals who wish to improve the quality of the nation's waters and to keep it that way for generations to come.

Richard L. Duesterhaus
Deputy Chief for Soil Science
and Resource Assessment
November 14, 1995

PREFACE

Since the *Water Quality Indicators Guide: Surface Waters* was first published six years ago, it has become for many users a good screening tool to inexpensively assess water quality potentials over large areas.

In fact, the demand for the Guide has grown far beyond its original intended audience of USDA field personnel to include educators, environmental professionals, local governments and water quality monitoring groups. Thus, the supply of these books has been long exhausted.

The Terrene Institute has worked with the Guide's co-author and NRCS' current national environmental coordinator, Charles Terrell, to update the Guide and produce this second edition in cooperation with Kendall/Hunt Publishing Company.

The Guide remains dedicated to an earlier NRCS national environmental coordinator, Vernon M. Hicks, whose vision of a tool to help NRCS field personnel easily and accurately recognize and remedy water quality problems evolved into this Guide. Mr. Hicks believed that environmental conditions could be surveyed without elaborate chemical testing procedures and water quality judgements made using environmental "surrogates" to represent pollution potential.

That concept developed into this Guide, which uses field sheets that employ environmental indicators to evaluate the environment. Practices are then recommended to alleviate agricultural nonpoint source pollution situations.

The *Water Quality Indicators Guide* has helped many people assess and make decisions about water quality over the past several years, and given them the confidence to address an area they once believed open only to "qualified professionals." May this second edition be even more widely used—by those of all levels of expertise—and continue to aid in improving the nation's waters.

CONTENTS

Introduction vi

Chapter 1 - Pollution Related to Agriculture 1

Chapter 2 - Water Quality Field Analysis 9

Chapter 3 - Ecology of Freshwater Systems 17

Chapter 4 - Sediment 19

Chapter 5 - Nutrients 23

Chapter 6 - Pesticides 29

Chapter 7 - Animal Wastes 33

Chapter 8 - Salts 37

Appendix A - Water Quality Procedures 43

Appendix B - Aquatic Organisms 48

Appendix C - Glossary 80

Appendix D - References 83

Appendix E - Conservation and Best Management Practices 86

Appendix F - Field Sheets 90

INTRODUCTION

The *Water Quality Indicators Guide* examines five major sources of agriculturally related nonpoint source pollution—sediment, nutrients, animal waste, pesticides and salts. Field sheets are provided to enable the user to assess surface water quality problems easily and accurately and to select appropriate remedial practices. The field sheet concept was adapted from a Wisconsin Department of Natural Resources methodology (ref. 1-1). The field sheets are completed in the field through onsite observations, rather than chemical or physical measurements. Conservation and best management practices (BMPs) are recommended to reduce or eliminate nonpoint source pollution originating from agricultural lands.

The field sheets are arranged in matrix format with environmental indicators given for sediment, animal wastes, nutrients, pesticides and salts. Each indicator is divided into descriptions of the environment from excellent to poor, and each description is given a weighted numerical ranking. The user matches the individual description with what is observed in the water or on the land. By totaling the individual rankings, a score is obtained indicating the potential for agricultural nonpoint source problems. Practices can be selected from the list to alleviate problem situations.

This type of approach may be sufficient in *some* instances to *confirm* that a particular nonpoint source pollution problem exists. In *other* instances, it may lead you to *suspect* a given pollutant, which can then be confirmed or denied by additional scientific analysis. When available, dissolved oxygen meters, salinity and conductivity meters, and field test kits may be used to supplement the *Water Quality Indicators Guide* field sheets. However, acceptable determinations can be made by using the field sheets without test kits or meters. When a particular nonpoint source pollutant is identified, the user of this guide is directed to possible solutions (conservation and best management practices), which are listed by number on the field sheets.

There are two types of field sheets: one type for receiving waters, including streams, rivers, lakes and ponds; and another type for use on agricultural lands draining into the receiving waters. Chapter 1 reviews the overall distribution of agricultural nonpoint source problems. Chapter 2 gives a history of the water quality indicators approach and some general limitations of the *Water Quality Indicators Guide: Surface Waters*. Instructions for the water-based "A" type field sheets and for the land-based "B" type field sheets are contained in chapter 2. Chapter 3 presents background ecological information about aquatic ecosystems, especially stream systems.

Chapters 4 through 8 discuss the five major pollutants—sediment, nutrients, pesticides, animal wastes and slats. These chapters discuss in detail the water quality indicators enumerated in the water-based "A" series of field sheets. The "B" field sheets are designed to assess the pollutant generation potential of a particular field or pasture and are completed in the same way as the "A" field sheets. As an aid, a glossary of terms appears in appendix C.

With practice, the user of this guide will quickly learn water quality assessment procedures by using the guide's field sheets, and the user's ability to assess water quality situations accurately will also increase.

Chapter 1

Pollution Related to Agriculture

Recent reports acknowledge that a principal water quality problem in our Nation is nonpoint source pollution. The U.S. Environmental Protection Agency defines nonpoint source (NPS) pollution as precipitation-driven stormwater runoff, generated by land-based activities, such as agriculture, construction, mining, and silviculture. Agricultural nonpoint sources are crop and animal production activities. These activities result in diffuse runoff, seepage, or percolation of pollutants from the land to surface and ground waters (ref. 1-2). Problems relating to agricultural nonpoint source pollution can be observed in the entire range of water bodies from estuaries to lakes and impoundments, to rivers, streams, and even farm ponds. Ground water is also vulnerable to pollution. Contaminated wells and drinking water supplies are now being identified.

In general, water quality problems result from five categories of agriculturally related nonpoint source pollution: sediment, nutrients, animal wastes, pesticides, and salts. Figure 1-1 shows the geographic potential for nonpoint source pollution of surface waters. The potential for agricultural nonpoint source pollution problems, according to SCS's Second Resources Conservation Act (RCA) Appraisal report (ref. 1-3), is shown in figures 1-2 through 1-7:

Figure 1-1

Composite Potential for Nonpoint Source Pollution of Surface Waters.

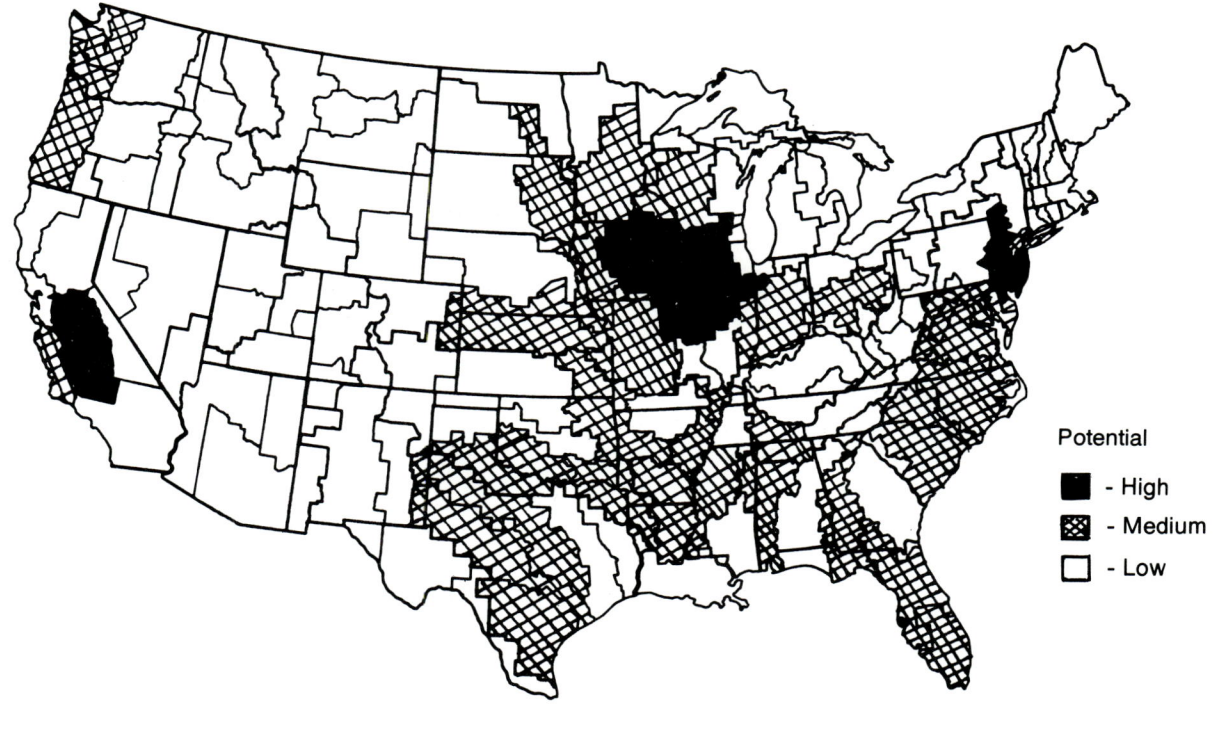

An area with a "low" composite rating could have a high rating for a specific contaminant. Ratings were made for multi-county watershed areas and do not identify more localized problems.

Figure 1-2

Potential for Pesticide Problems.

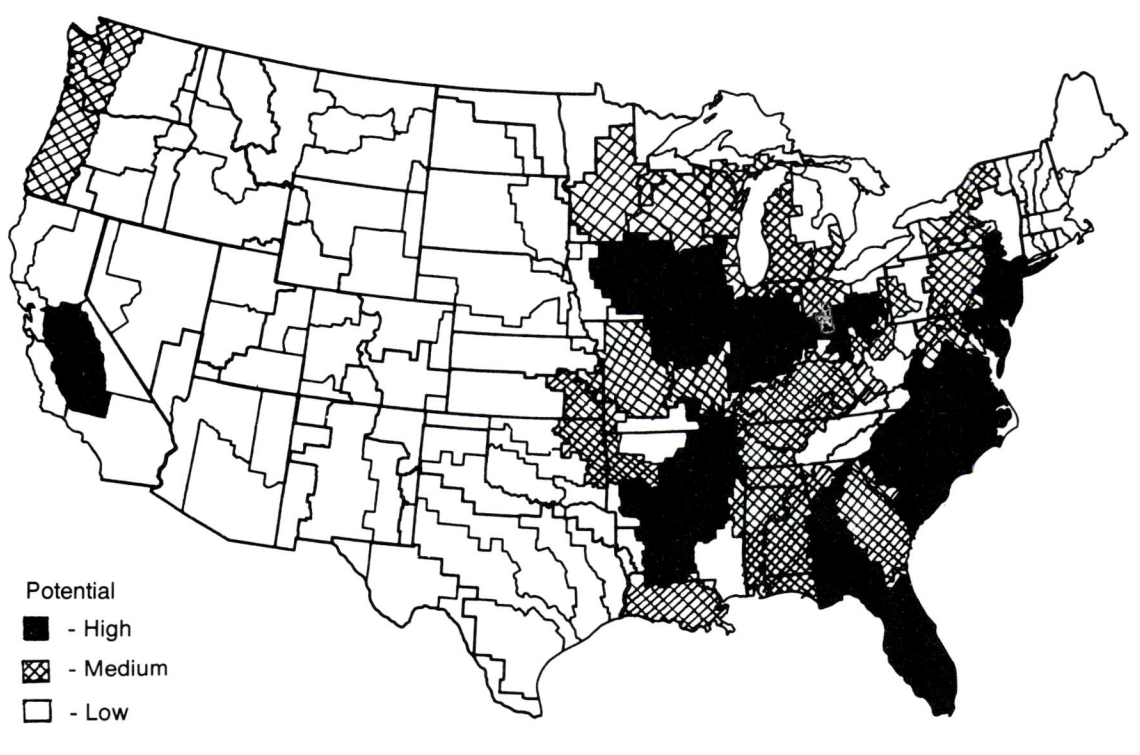

The potential for surface water pollution by pesticides was estimated by multiplying the crop acreages in each area by pesticide application coefficients for 184 pesticides. These values were multiplied by an availability factor that estimated the percentage of an application leaving a field and were adjusted by a runoff value for the growing season. Pollution potential is estimated for each watershed as a whole; localized conditions may be masked by aggregation. To confirm the existence of pesticide pollution, stream and lake monitoring would be necessary.

Figure 1-3

Tons of Manure Per Acre of Cropland and Grassland.

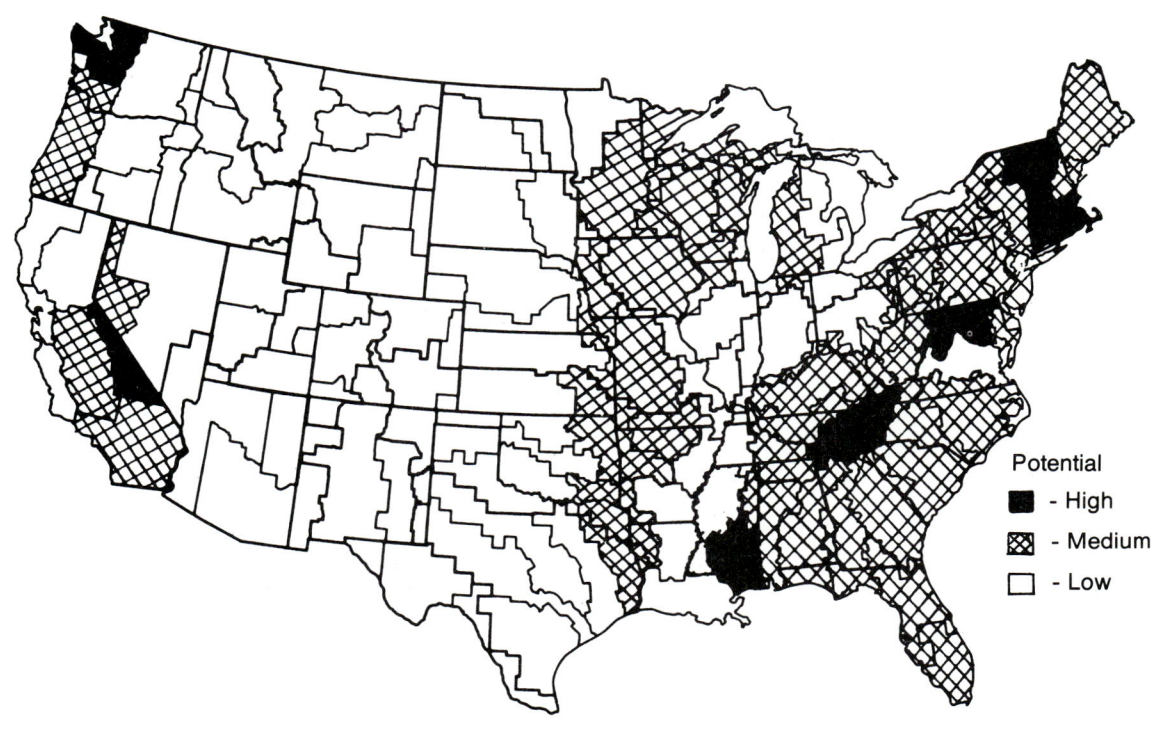

The number of each type of animal in a county (from the 1982 Agricultural Census) was multiplied by the appropriate manure production factor. The amounts of manure produced by all the county's livestock were totaled and aggregated by area; the total was divided by the acreage of cropland plus grassland (from the Agricultural Census) in each area.

Figure 1-4

Potential for Animal Waste Problems.

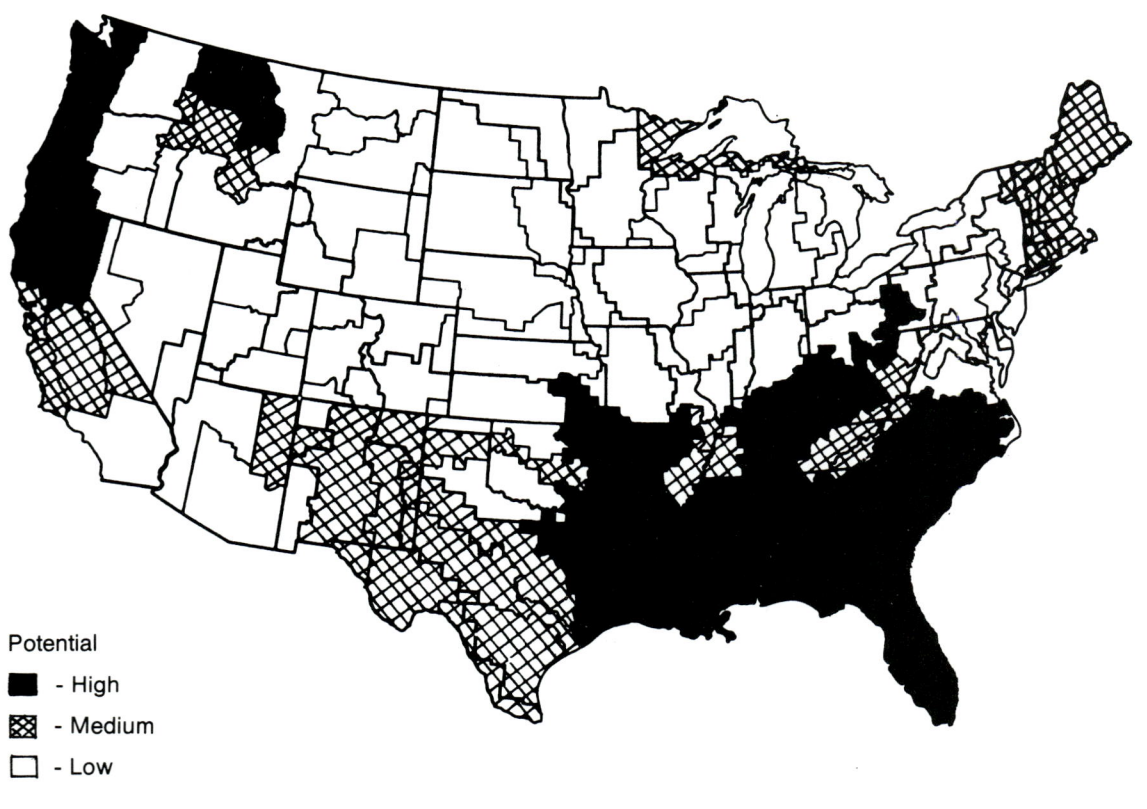

The figure shows potential for pollution resulting from animal wastes, taking into account percentage of manure needing improved management, percentage of cropland and grassland associated with animal enterprises, runoff from precipitation, ratio of feed purchased to feed produced on farm, and ratio of nitrogen and phosphorus available from manure to nitrogen and phosphorus needed by crops.

Figure 1-5

Potential for Nutrient Problems.

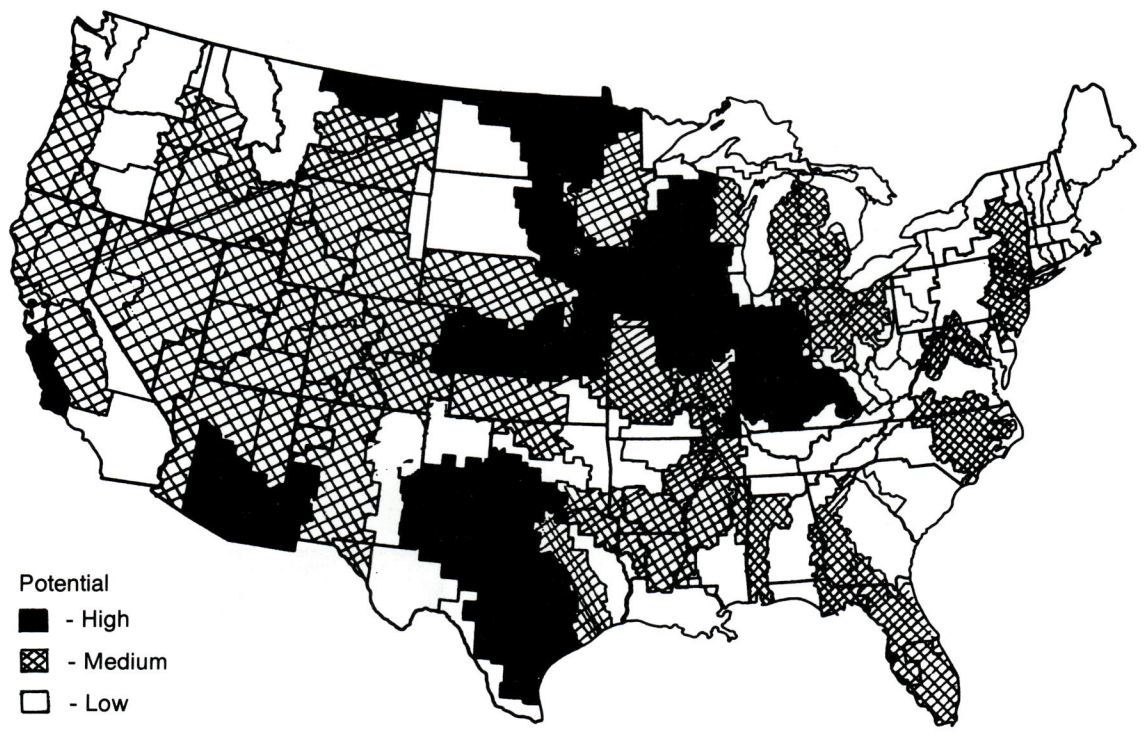

Source: WATSTORE (U.S. Geological Survey data from water quality stations, Ref. 1-4).

The potential for impairment of water quality was estimated by determining nutrient concentrations, by form, and comparing them with the respective threshold levels at which they threaten desired water uses. Data on nutrient concentrations were taken from WATSTORE (U.S. Geological Survey data from water quality stations). Stations were primarily National Stream Quality Accounting Network stations at the downstream and of hydrologic accounting units. Estimates of pollution potential are for the watershed as a whole and may not reflect localized conditions.

Figure 1-6

Estimated Sediment Yield.

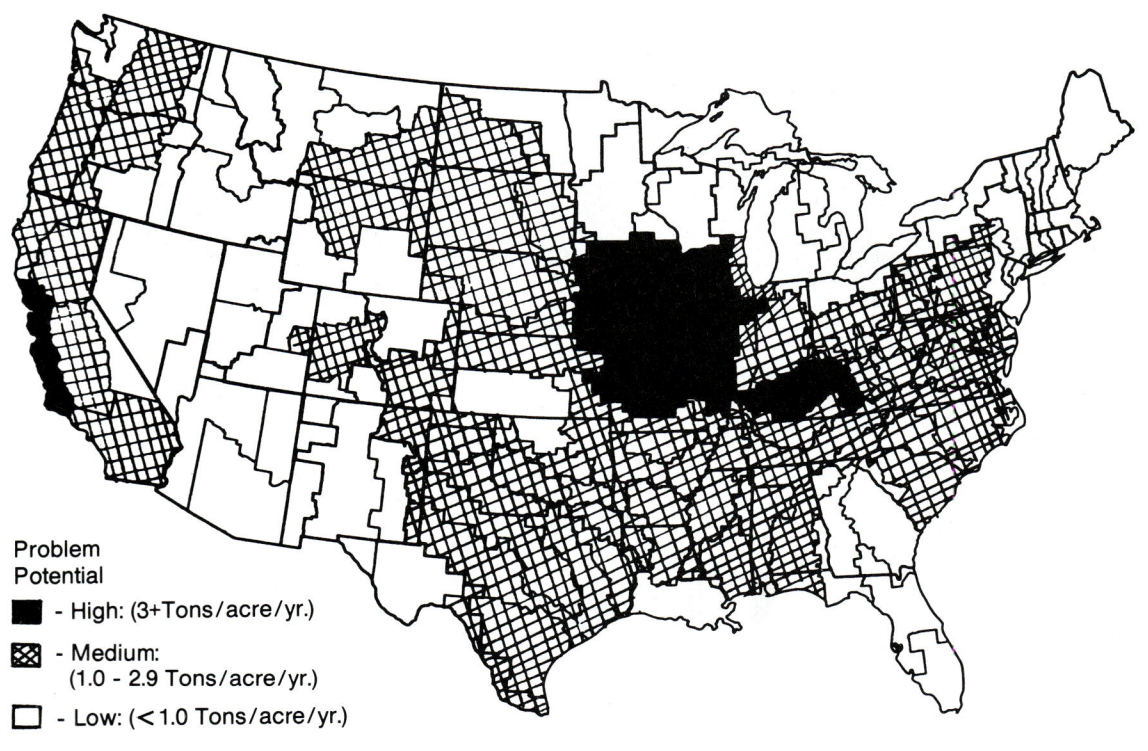

Sources: (1) 1982 National Resources Inventory (USDA-SCS, 1984, Ref. 1-5).
(2) USGS Surface Soil Surveys (Ref. 1-6).
(3) USDA Soil Survey Laboratory Data State Reports (Ref. 1-7).

Estimated sheet and rill erosion rates reported in the 1982 NRI were adjusted to county boundaries. Sediment delivery for each county and land use was estimated using state sediment delivery curves developed for the 1977 NRI. Sediment delivery rates are assumed to be higher in areas where streams are more numerous and closely spaced and where the surface soils have a higher percentage of fine particles (silt and clay). Data from USGS Surface Soil Surveys and USDA Soil Survey laboratory data were analyzed also.

Figure 1-7

Potential for Salinity Problems.

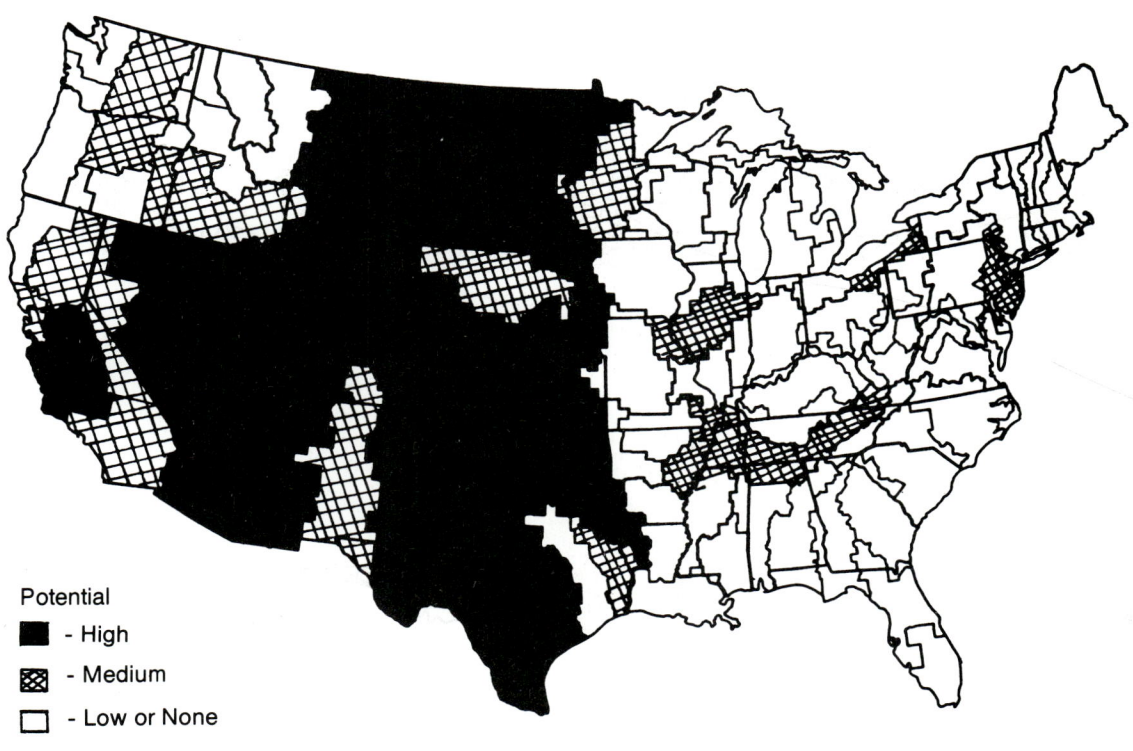

Sources: (1) U.S. Geological Survey National Stream Quality Accounting Network (NASQAN) stations in ASAs (Ref. 1-8).
(2) Published and unpublished data from EPA and USGS.

To assess potential, indicators of total dissolved solids, adjusted sodium adsorption, and chloride concentration were checked and total solid loads were analyzed using data for agricultural acreages, areas affected by saline or sodic soils, and irrigated acres as modifying and/or contributing factors. Data analyzed were taken from the U.S. Geological Survey National Stream Quality Accounting Network stations and published and unpublished data from EPA and USGS.

Chapter 2

Water Quality Field Analysis

History of the Indicators Approach

Two centuries ago, when the U.S. population was small, the number of farmers and farm animals was also small. Agricultural wastes did not overload streams or other receiving water bodies. In those days, streams cleansed themselves naturally. Today, with the increasing complexity of farms, many watercourses and water bodies are unable to cope with the pollution loads being generated.

The SCS *Water Quality Indicators Guide: Surface Waters* is designed to determine by means of an indicators approach whether farm-generated materials are a problem. Water pollution investigators have used this type of approach since the turn of the century. At the heart of this approach is a comparison of water quality conditions above and below a suspected source of pollution. In most instances, the suspected source may be a "point" source pollution; that is, a type of pollution that can be readily identified as coming from a discrete source, such as a discharging pipe (e.g., a sewage outfall).

The *Water Quality Indicators Guide* adapts this approach for use with nonpoint source pollution—pollutants whose sources are diffuse and not readily identifiable. Nonpoint source pollutants include those substances which run off, wash off, or seep through the ground into receiving watercourses and water bodies. Agricultural nonpoint source pollution tends to wash or run off large tracts of cropland, pastures, feedlots, etc., and the conditions leading to pollution are highly variable.

One of the most important pollution variables is flow. In nonirrigated regions, loadings of the most common nonpoint source pollutants in a small stream tend to be proportional to the amount of runoff. Runoff, in turn, varies with conditions, such as: (1) amount of snowmelt or rainfall; (2) rate of snowmelt or rainfall; (3) soil type, condition, slope, vegetative cover, and land use; (4) time elapsed since the previous storm; and (5) seasonal timing and intensity of storm events.

Not only are the timing and extent of nonpoint source pollution events highly variable, but the effects of nonpoint source pollutants, either singly or in combination, are also variable. The effect of a given pollutant on water quality depends upon local site-specific environmental conditions; that is, on the local geology and the physical/chemical characteristics of the nearby water.

Both water quality and rate of flow influence the types of organisms that inhabit a given watercourse or water body. Organisms respond to many local environmental conditions, including climate, habitat availability, streambed type, etc. The ecology of watercourses is discussed in the next chapter.

Limitations of the Water Quality Indicators Guide

The *Water Quality Indicators Guide* was written to cover the entire United States, so it is general by intent. It can be expected that a particular stream or pond may deviate from the norms presented and will require the user to make adjustments for local situations. However, the guide has been field tested in five States across the Nation and by individual Soil Conservation Service personnel from many other States. The ideas, suggestions, and comments from those tests have been incorporated into this version. The *Indicators Guide* is not a research tool, nor does it give quantitative data, but as a qualitative tool and as an educational or learning device, it will aid the user in evaluating agricultural nonpoint source pollution problems.

This guide is especially limited where water flow rates are excessively low or high. In ephemeral or intermittent streams, some parts of this guide, such as observing fish, vegetation, or bottom invertebrates, cannot be used. The guide's use may be limited in heavily silted, mud-bottom streams, where the silt's presence provides an unsuitable habitat for many species. Also, heavy siltation of the water can "mask" the effects of nutrients that may be present, by shutting out light that normally would reach aquatic vegetation, allowing its growth. Thus, the vegetational part of the nutrient field sheet may not work well in heavily silted waters. In these cases, chemical testing may be necessary to determine nutrient levels.

Description of the Field Sheets

The heart of the *SCS Water Quality Indicators Guide: Surface Waters* is a series of field sheets (appendix F). The field sheets relate to surface water quality and are designed to help field personnel assess the degree of contribution to receiving waters from agriculturally related pollutants, namely sediment, animal waste, nutrients, pesticides, and salts. The receiving watercourses are natural streams, constructed channels, or receiving water bodies, such as ponds or lakes.

The field sheets are of two types: "A" and "B." The five "A" field sheets are designed to assess the effects of pollutants to receiving waters. These are water-based field sheets and should be completed onsite, following visual inspection of the receiving water.

By contrast, the seven "B" field sheets are land-based and are designed to assess the pollutant potential of a particular field or pasture; i.e., how likely it is that an agriculturally produced pollutant will be carried from a given field or pasture to a receiving watercourse or water body, or to ground water. There are more "B" field sheets than "A" sheets, because some land-based activities or environmental conditions required special emphasis.

Procedure for Field Analysis

NOTE: **Do not** write on the original field sheets. Make a copy of each field sheet before proceeding and write on the copies.

Step 1. Begin by completing the background information section (part 1) of the "Watershed Assessment." Although the Watershed Assessment was designed to be used with natural perennial streams, it can be adapted for use on either intermittent or ephemeral streams or on constructed waterways.

Please note that this evaluation cannot be made in the office. It must be made onsite, in the field. If you lack some of the necessary information, seek it from the landowner or operator, county agricultural extension agent, biologist, or other knowledgeable person.

Step 2. The "On-Farm (Ranch) Water Assessment" should be completed for each farm or ranch visited.

Step 3. Next, do a preliminary assessment of possible nonpoint source impacts by answering the questions asked in the "Watercourses" or "Water Bodies" Field Sheet Selection (part 2). If any of the questions in part 2 of the assessment receives a "yes" answer, then it is likely that the receiving water is being adversely affected by the pollutant indicated in the last column under the heading "Probable Cause." You can verify this by completing the field sheets for this particular pollutant.

Please note that it is much easier to determine nonpoint source (NPS) pollution effects on standing (lentic) water, such as lakes or ponds, than for flowing (lotic) water, because standing water has a longer residence time (time that water remains in the water body), giving pollutants time to react.

Step 4. Proceed to the field sheets. If you are confident of your "no" answers in part 2 of the above assessment, you need to complete only those field sheets corresponding to the questions (pollutants) for which you marked either a "yes" or "can't tell" answer. For example, there will not be an animal waste problem if a particular farm or ranch has no animals and the owner or operator does not import animal waste. Obviously, in this case, none of the animal waste field sheets (2A, $2B_1$, $2B_2$) needs to be completed. If you are not confident that any of the pollutants should be eliminated as possible contributors of NPS pollution in a particular situation, complete all of the field sheets.

To learn how to use the sheets, it is recommended that you go through all of them at least once, including those for pollutants that have just a small possibility of affecting the watercourse or water body. This will allow you to gain familiarity with the sheets. With practice, using the sheets will become second nature to you, and you will complete them very quickly.

Filling Out the Field Sheets

TYPE A FIELD SHEETS

If upon completing part 2 of the watercourse (or water body) assessment you determined that sediment is probably adversely affecting the water, you should begin by focusing on the water-based Field Sheet 1A: "Sediment Indicators for Receiving Watercourses and Water Bodies (fig. 2-1)." Please take time now to look at this sheet. Outlined below is how you should use it. The sheet has answers circled in the way that should be done in the field.

For each field sheet, you are asked to complete the blanks at the top of the sheet which identify you, the evaluator, the county, State, etc. Notice that in the left column, Field Sheet 1A lists six different indicators or rating items with four possible options for item number 3. You will examine one indicator at a time and judge whether the water quality at this particular site ranks as excellent, good, fair, or poor regarding that particular indicator. Please note that these sheets should be completed in the field at the water's edge and *not* in the office.

A standing water body is fairly easy to assess for nonpoint source pollutant impacts. Flowing waters are not as easy to evaluate. The best place to observe a receiving watercourse is downstream of the pollutant sources. The exact point downstream from which to observe varies. If the water flow is very rapid, you may have to make observations at a distance downstream where the flow is slower. This is especially true when using the Nutrient Field Sheet (3A) because the effects from excessive nutrients often do not show in flowing waters until the flow rate is slow.

In completing Field Sheet 1A, it would be best to station yourself beside the stream (fig. 2-2) at the spot indicated by the *A. If the stream is flowing rapidly, flushing away pollutants very quickly, it may be necessary to walk downstream or upstream, observing indicators as you go. For ponds and lakes, it is best to observe from a site that allows a bird's-eye view of the whole water body, as well as from the water's edge.

The first indicator or ranking item on Field Sheet 1A for sediment is turbidity. Note that an indication of nonpoint source sediment pollution can most accurately be assessed only during or immediately following a storm event. Ask yourself, "What does the water look like at this particular site immediately after a storm?" Do you see "conditions normally expected under pristine conditions in your geographic region?" Is the water "clear or very slightly muddy after a storm event" or are "objects visible at depths greater than 3 to 6 feet (depending on water color)," such as described under the EXCELLENT heading? Or do the descriptors under the GOOD category more closely approximate conditions in your area; i.e., the water is "what is expected for *properly* managed agricultural land in your geographic region?" Is the water "a little muddy after a storm event but clears rapidly" or are "objects visible at depths between 1-1/2 to 3 feet (depending on water color)?" Are the conditions at this site better described by the descriptors under the headings of FAIR or POOR? Having read all four definitions under each of the four ratings, decide which of the four BEST describes the condition of the watercourse or body which you are evaluating and circle the number in the bottom of the box for that particular rating.

Follow the procedure outlined above for the turbidity parameter with each of the other five rating items on the Sediment Field Sheet 1A. When you have completed the entire sheet, add the circled numbers to obtain a total for the entire field sheet. This total should fall into one of the four ranking categories (excellent, good, fair, or poor) given at the very bottom of each field sheet. For example, if the total score was "8," record an "8/Poor" in the upper right-hand corner of the field sheet by "Total Score/Rank." What this says is that the water being evaluated is in a "poor" condition relative to sediment—or that sediment is greatly impacting the water at this site.

Design and Tailoring of the Indicator Guide Field Sheets To Fit Your Region

Please note that the field sheets are designed to be used for both flowing water and standing water across the entire United States. To use the sheets throughout this exceedingly diverse geographic area and for flowing and standing waters, it was necessary to include several descriptors per indicator (rating item) in each of the four categories (excellent, good, fair, and poor). These descriptors will rarely fit *all* given situations in a particular geographic area. In fact, some of the options within the same rank might at first appear contradictory if you fail to distinguish between standing and flowing water. Be especially careful when reading these descriptors and be sure to select the option which BEST or most closely matches the site specific conditions of the water you are assessing.

If the condition of the water in your locality *really* falls between two options or has about half of the characteristics of two options, you may "split" a score. You may want to add one or two other descriptors to all four options of a rating item. These

Figure 2-1

Sediment Page 1 of 2

FIELD SHEET 1A: SEDIMENT
INDICATORS FOR RECEIVING WATERCOURSES AND WATER BODIES

Lat. 40° 37' 30"
Lon. 76° 40' 00"

Evaluator: **Isaacs/Myers** County/State: **Dauphin, PA** Date: **18 Apr. '88**
Water Body Evaluated: **Pond** Water Body Location: **Lykens, PA** Total Score/Rank: **24 - Good**

Rating Item	Excellent	Good	Fair	Poor

(Circle one number among the four choices in each row which BEST describes the conditions of the watercourse or water body being evaluated. If a condition has characteristics of two categories, you can "split" a score.)

1. Turbidity (best observed immediately following a storm event)

- Excellent:
  - What is expected under pristine conditions in your region.
  - Clear or very slightly muddy after storm event.
  - Objects visible at depths greater than 3 to 6 ft. (depending on water color).
  - OTHER
  - **9**

- Good:
  - What is expected for properly managed agricultural land in your region.
  - A little muddy after storm event but clears rapidly.
  - Objects visible at depths between 1½ to 3 ft. (depending on water color).
  - OTHER
  - **(7)**

- Fair:
  - A considerable increase in turbidity for your region.
  - Considerable muddiness after a storm event. Stays slightly muddy most of the time.
  - Objects visible to depths of ½ to 1½ ft. (depending on water color).
  - OTHER
  - **3**

- Poor:
  - A significant increase in turbidity for your region.
  - Very muddy—sediment stays suspended most of the time.
  - Objects visible to depths less than ½ ft. (depending on water color).
  - OTHER
  - **0**

2. Bank stability in your viewing area

- Excellent:
  - Bank stabilized.
  - No bank sloughing.
  - Bank armored with vegetation, roots, brush, grass, etc.
  - No exposed tree roots.
  - OTHER
  - **10**

- Good:
  - Some bank instability.
  - Occasional sloughing.
  - Bank well-vegetated.
  - Some exposed tree roots.
  - OTHER
  - **(7)**

- Fair:
  - Bank instability common.
  - Sloughing common.
  - Bank sparsely vegetated.
  - Many exposed tree roots & some fallen trees or missing fence corners, etc.
  - Channel cross-section becomes more U-shaped as opposed to V-shaped.
  - OTHER
  - **4**

- Poor:
  - Significant bank instability.
  - Massive sloughing.
  - No vegetation on bank.
  - Many fallen trees, eroded culverts, downed fences, etc.
  - Channel cross-section is U-shaped and stream course or gully may be meandering.
  - OTHER
  - **1**

3. Deposition (Circle a number in only A, B, C, or D)

SELECT 3A OR 3B OR 3C OR 3D

3A. Rock or gravel streams OR

- Excellent:
  - A. For rock and gravel bottom streams:
  - Less than 10% burial of gravels, cobbles, and rocks.
  - Pools essentially sediment free.
  - **9**

- Good:
  - A. For rock and gravel bottom streams:
  - Between 10% & 25% burial of gravels, cobbles, & rocks.
  - Pools with light dusting of sediment.
  - **7**

- Fair:
  - A. For rock & gravel bottom streams:
  - Between 25% and 50% burial of gravels, cobbles and rock.
  - Pools with a heavy coating of sediment.
  - **3**

- Poor:
  - A. For rock & gravel bottom streams:
  - Greater than 50% burial of gravels, cobbles and rocks.
  - Few if any deep pools present.
  - **1**

3B. Sandy bottom streams OR

- Excellent:
  - B. For sandy streambeds:
  - Sand bars stable and completely vegetated.
  - No mudcaps or "drapes" (coverings of fine mud).
  - No mud plastering of banks; exposed parent material.
  - No deltas.
  - **9**

- Good:
  - B. For sandy streambeds:
  - Sand bars essentially stable and well, but not completely, vegetated.
  - Occasional mudcaps or "drapes."
  - Some mud plastering of banks.
  - Beginnings of delta formation.
  - **7**

- Fair:
  - B. For sandy streambeds:
  - Sand bars unstable with sparse vegetation.
  - Mudcaps or "drapes" common.
  - Considerable mud plastering of banks.
  - Significant delta formation.
  - **3**

- Poor:
  - B. For sandy streambeds:
  - Sand bars unstable and actively moving with no vegetation.
  - Extensive mudcaps or "drapes."
  - Extensive mud plastering of banks.
  - Extensive deltas.
  - **1**

3C. Mud-bottom streams OR

- Excellent:
  - C. For mud bottom streams:
  - Dark brown/black tanic-colored water (due to presence of lignins and tanins).
  - Abundant emergent rooted aquatics or floating vegetation.
  - **9**

- Good:
  - C. For mud bottom streams:
  - Dark brown colored water.
  - **7**

- Fair:
  - C. For mud bottom streams:
  - Medium brown water, muddy bottom.
  - **3**

- Poor:
  - C. For mud bottom streams:
  - Light brown colored, very muddy bottom.
  - **1**

11

Figure 2-1

Sediment Page 2 of 2

FIELD SHEET 1A: SEDIMENT, Continued
INDICATORS FOR RECEIVING WATERCOURSES AND WATER BODIES

Rating Item	Excellent	Good	Fair	Poor
3D. Ponds	:-- Ponds essentially sediment free. :-- No reduction in pond storage capacity. :-- OTHER 9	:-- Ponds with light dusting of sediment. :-- Very little loss in pond storage capacity. :-- OTHER 7	:-- Ponds with a heavy coating of sediment. :-- Some measurable loss in pond storage capacity. :-- OTHER (3)	:-- Ponds filled with sediment. :-- Significant reduction in pool storage capacity. :-- OTHER 1
4. Type and amount of aquatic vegetation & condition of periphyton (plants, growing on other plants, twigs, stones, etc.)	:-- Periphyton bright green to black. Robust. :-- Abundant emergent rooted aquatics or shoreline vegetation. :-- In ponds, emergent rooted aquatics (e.g. cattails, arrowhead, pickerelweed, etc.) present, but in localized patches. :-- OTHER 9	:-- Periphyton pale green and spindly. :-- Emergent rooted aquatics or shoreline vegetation common. :-- In ponds, emergent rooted aquatics common, but confined to well-defined band along shore. :-- OTHER (7)	:-- Periphyton very light colored or brownish and significantly dwarfed. :-- Sparse vegetation. :-- In ponds, emergent rooted aquatics abundant in wide band; encroachment of dry land species (grasses, etc.) along shore. :-- OTHER 5	:-- No periphyton. :-- No vegetation. :-- In ponds, emergent rooted aquatics predominant with heavy encroachment of dry land species. :-- OTHER 2
OPTIONAL: 5. Bottom stability of streams	:-- Stable. :-- Less than 5% of stream reach has evidence of scouring or silting. :-- OTHER 9	:-- Slight fluctuation of streambed up or down (aggradation or degradation). :-- Between 5-30% of stream reach has evidence of scouring or silting. :-- OTHER 7	:-- Considerable fluctuation of streambed up or down (aggradation or degradation). :-- Scoured or silted areas covering 30-50% of evaluated stream reach. :-- Flooding more common than usual. :-- More stream braiding than usual for region. :-- OTHER 3	:-- Significant fluctuation of streambed up or down (aggradation or degradation). :-- More than 50% of stream reach affected by scouring or deposition. :-- Flooding very common. :-- Significantly more stream braiding than usual for region. :-- OTHER 1
OPTIONAL: 6. Bottom dwelling aquatic organisms	:-- Intolerant species occur: mayflies, stoneflies, caddisflies, water penny, riffle beetle and a mix of tolerants. :-- High diversity. :-- OTHER 9	:-- A mix of tolerants: shrimp, damselflies, dragonflies, black flies. :-- Intolerants rare. :-- Moderate diversity. :-- OTHER 7	:-- Many tolerants (snails, shrimp, damselflies, dragon flies, black flies). :-- Mainly tolerants and some very tolerants. :-- Intolerants rare. :-- Reduced diversity with occasional upsurges of tolerants, e.g. tube worms and chironomids. :-- OTHER 3	:-- Only tolerants or very tolerants: midges, craneflies, horseflies, rat-tailed maggots, or none at all. :-- Very reduced diversity; upsurges of very tolerants common. :-- OTHER 1

1. Add the circled Rating Item scores to get a total for the field sheet.   TOTAL [**24**]
2. Check the ranking for this site based on the total field score. (Check "excellent" if the score totals at least 32. Check "good" if the score falls between 21 and 31, etc.). Record your total score and rank (excellent, good, etc.) in the upper right-hand corner of the field sheet. If a Rating Item is "fair" or "poor," complete Field Sheet 1B.

RANKING	Excellent (32-37) [ ]	Good (21-31) [**24**]	Fair (9-20) [ ]	Poor (8 or less) [ ]
OPTIONAL RANKING (with #5 OR #6)	Excellent (40-46) [ ]	Good (26-39) [ ]	Fair (11-25) [ ]	Poor (10 or less) [ ]
OPTIONAL RANKING (with #5 AND #6)	Excellent (48-55) [ ]	Good (31-47) [ ]	Fair (13-30) [ ]	Poor (12 or less) [ ]

Figure 2-2

Nonpoint Source Pollution Effects.

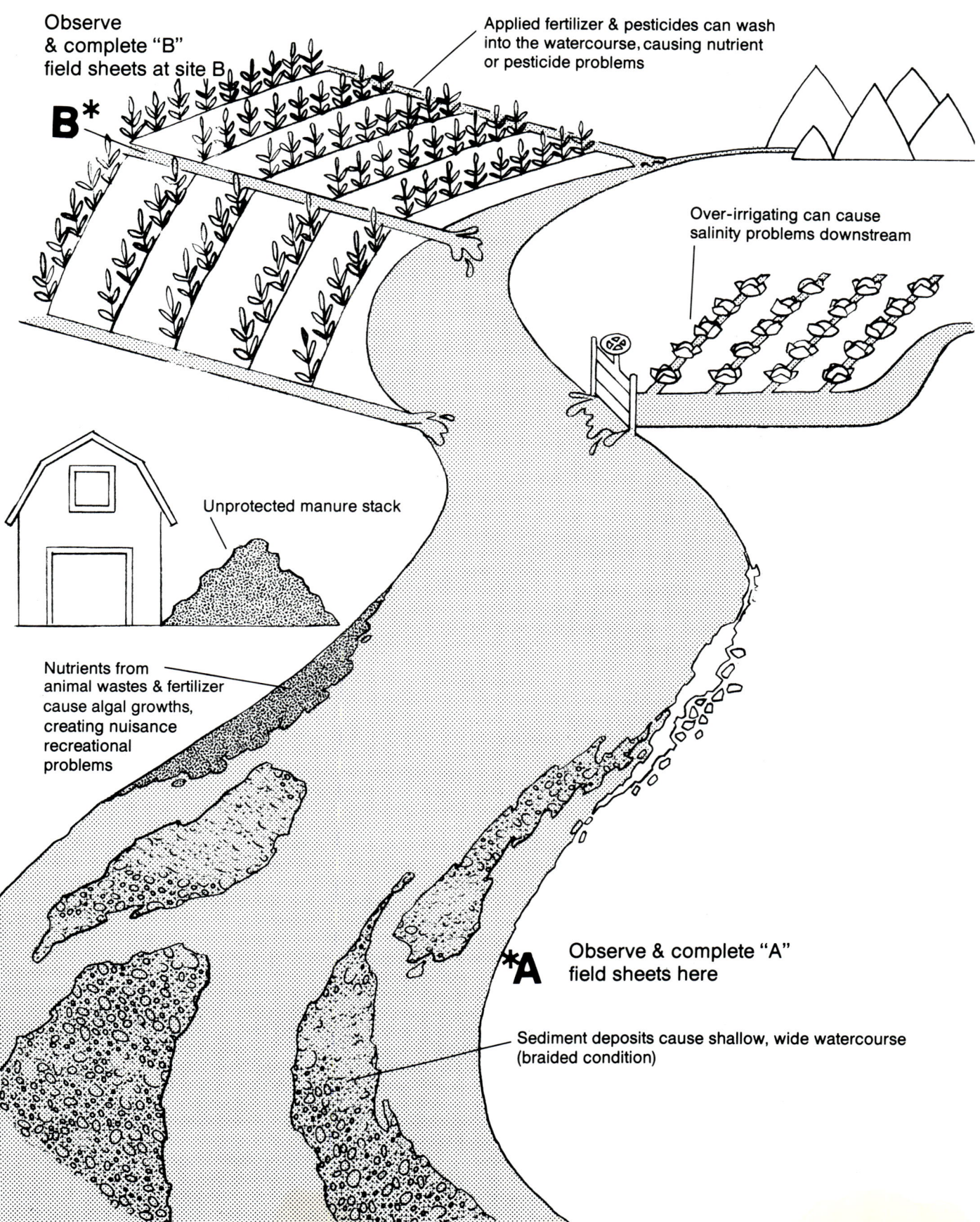

other options apply to your particular geographic region and precisely define particular water quality situations. The word "OTHER" that has been included in each block on each field sheet means that you are free to adapt the field sheets to your particular region or locale. Note also that if none of the descriptors fit, you can resort to rankings relative to your geographic region, such as the first ones given for the turbidity indicator on Field Sheet 1A: Sediment.

One last point—A field sheet, like any other tool or instrument, is only as good as the person using it. This is true of the use of these field sheets. Those who take the time to learn how to use the *Water Quality Indicators Guide* field sheets will quickly become proficient in their use. Based on your experience with the sheets, you will start to make judgments about water quality and will develop an "intuitive feel" for the water's condition. *Rely on this judgment,* even if it means altering the field sheets.

Remember that the field sheets are only as good a tool as you make them, especially concerning local conditions.

Given that water is severely polluted by sediment, how can we know that the sediment is coming from agriculturally related activities? If it is related to agriculture, how can we correct the problem and improve water quality? To answer these questions, turn to the "B" field sheets.

TYPE B FIELD SHEETS

Assumption. Before using the series "B" field sheets, it is important to recognize that underlying the design of the overall field analysis is the assumption that we are striving for water of fishable/swimmable qualities—a goal established in the Federal Water Pollution Control Act of 1972 and iterated in the 1987 Water Quality Act Amendments. While geographic and site-specific conditions might cause us to accept a "good" rating in some instances, we should not be satisfied with a water quality rating of "fair" or "poor."

The "B" field sheets should be completed in all cases where water quality ranks lower than what is expected regionally under naturally occurring pristine conditions for any of the five major agricultural pollutants. While in many cases the pristine condition will receive an excellent rating, in other cases naturally occurring conditions (geologic, topographic, etc.) prevent the waters from ever being "excellent" (fishable/swimmable). It is important to be able to distinguish between naturally occurring and human-induced limitations to water use. It may be difficult to determine what constitutes "pristine" conditions for your area. If you do not know or are not sure, be sure to consult with local experts in the water quality field. Call the SCS State Office Water Quality Specialist or Biologist or the specialists at the SCS National Technical Centers. Every State has a water pollution control agency, although the names vary.

Specialists in these offices are most willing to assist. Additionally, many local colleges and universities have environmental and water quality experts who can be of great help.

The "B" field sheets allow an on-farm or on-ranch assessment (fig. 2-3) of the five major agriculturally related contributors of pollution. Recommendations for improving problem situations are given in the last column of each sheet under "Practices from appendix E" (conservation and best management practices, ref. B-6). Figure 2-3 is completed with circled answers in the way that was done on Field Sheet 1A.

Other than the list of conservation and best management practices (BMP's) in the last column, the format for the "B" series is identical to that for the "A" series. Therefore, the procedure outlined above for use with the "A" field sheets should also be used in completing the "B" sheets.

The "B" sheets should be completed onsite. If a conservation plan exists for a given property, it would be helpful to have it in hand while completing the "B" field sheets. A soil survey of the area would also be helpful if you are not familiar with the land tract. You may want to briefly reconnoiter the tract of land. Previous experience with this particular property owner or manager and prior knowledge of the property will prove invaluable.

Based on your previous knowledge of the land or your recent reconnaissance, define a "representative" field which drains into a watercourse or water body you have judged to be polluted by use of the "A" field sheets. That is, choose an area large enough to give an appropriate numerical weighting to both properly and poorly managed areas. Then proceed to complete the appropriate "B" field sheet relative to the field that you just defined. While a sample field size should be representative, it is recommended to select for your observation site a location where you could expect to find a pollutant. For example, if you were assessing nutrients or pesticides, you might stand in the middle of the row crops, as shown in figure 2-2, where the B* is indicated. If you were interested in sediment pollution, you might position yourself in or near a recently plowed field.

If scores for any of the indicators (rating items) were ranked less than "good" or "excellent," you will want to consider recommending to the property owner or user one or more of the conservation or BMP's listed in the right-hand column of the sheet for that particular rating item. The practices listed are by no means exhaustive and may not be entirely suitable to your locality. Therefore, you will need to evaluate the suggested practices, selecting those that you consider to be appropriate to the given situation and adding others that may be lacking.

Figure 2-3

Sediment

FIELD SHEET 1B: SEDIMENT
INDICATORS FOR CROPLAND, HAYLAND OR PASTURE

Lat. 40° 37' 30"
Lon. 76° 40' 00"

Evaluator: **Isaacs/Myers** County/State: **Dauphin, PA** Date: **18 Apr. '88**
Field Evaluated: **Wilson** Field Location: **Lykens, PA** Total Score/Rank: **30 Good**

Practices from Appendix E

(Circle one number among the four choices in each row which BEST describes the conditions of the field or area being evaluated. If a condition has characteristics of two categories, you can "split" a score.)

Rating Item	Excellent	Good	Fair	Poor	Practices from Appendix E
1. Erosion Potential	-- Not significant. -- Less than T (tolerance); little sheet, rill, or furrow erosion. -- No gullies. -- OTHER **10**	-- Some erosion evident. -- About T; some sheet, rill, or furrow erosion. -- Very few gullies. -- OTHER **(7)**	-- Moderate erosion. -- T to 2T. -- Gullies or furrows from heavy storm events obvious. -- OTHER **3**	-- Heavy erosion. -- More than 2T. -- Many gullies or furrows & presence of critical erosion areas. -- OTHER **0**	1,3,5,7,8, 9,10,11, 15,16,17, 18,19,20, 21,22,23, 24,25,26, 27,29,30, 31,32,33, 37,38,40, 45,46,54, 61,62,65, 69,70,73, 75,79,85, 87,95,97, 99,102
2. Runoff Potential	Low: -- Very flat to flat terrain (0-0.5% slope). -- Runoff curve number (RCN) 61 - 70. -- Dry, low rainfall (less than 20"). -- Even, gentle impact (scattered shower-type) rainfall. -- OTHER **10**	Moderate: -- Flat to gently sloping (0.5-2.0% slope). -- RCN 71 - 80. -- Semidry (20-30"). -- Even, gentle to moderate intensity rainfall. -- OTHER **(8)**	Considerable: -- Gently to moderately sloping (2.0-5.0% slope). -- RCN 81 - 90. -- Semiwet (30-40"). -- Even to uneven intense rainfall. -- OTHER **4**	High: -- Moderately sloping to steep terrain (greater than 5%). -- RCN greater than 90. -- Wet (more than 40"). -- Intense uneven rainfall, especially in seasons when soil is exposed. -- OTHER **0**	6,9,88,95
3. Filtering effect or sedimentation potential of a vegetated buffer or water/sediment collecting basin	-- Intervening vegetation between cropland & watercourse greater than 200 ft. -- Type of intervening vegetation ungrazed woodland, brush, or herbaceous plants. -- Water & sediment control basins properly installed & maintained. -- OTHER **8**	-- Intervening vegetation between cropland & watercourse 100 to 200 ft. -- Type of intervening vegetation grazed woodland, brush, or herbaceous plants or range. -- Water & sediment control basins properly installed but poorly maintained. -- OTHER **6**	-- Intervening vegetation between cropland & watercourse 50 to 100 ft. -- Type of intervening vegetation high density cropland. -- Water & sediment control basins poorly installed & poorly maintained. -- OTHER **(4)**	-- Cropping from less than 50 ft up to water's edge. -- Type of intervening vegetation low density cropland or bare soil. -- No water & sediment control basins. -- OTHER **2**	5,18,25, 27,79,107
4. Resource management systems (RMS's) on whole farm (combined value for all agricultural areas	-- Excellent management. -- RMS's always used as needed. -- OTHER **9**	-- Good management. -- Most (80%) of the needed RMS's installed. -- OTHER **(7)**	-- Fair management. -- About 50% of the needed RMS's installed. -- Cropping confined to proper land class. -- OTHER **3**	-- Poor management. -- Few, if any, needed RMS's installed. -- Cropping not confined to proper classes. -- OTHER **0**	Practices same as Rating Item #1
5. Potential for ground water contamination	LOW: -- Soils rich to very rich in organic matter (greater than 3.0%). -- Slow to very slow percolation in light textured soils such as clays, silty or sandy clays, or silty clay loams. -- Perched water table present. -- In protected bedrock areas (50 ft. of soil & shale cap), well depth is 75-100 ft. -- In protected bedrock areas overlain with 50 ft. of sand or gravel, well depth is greater than 150 ft. -- In shallow bedrock areas (25-50 ft. soil & shale cap), well depth greater than 200 ft. -- In Karst areas, well depth is greater than 1,000 ft., if aquifier is "confined." -- OTHER **9**	MODERATE: -- Soils rich to moderate in organic matter (3.0 to 1.5%). -- Slow to moderate percolation in clay loams or silts. -- Perched water table present. -- In protected bedrock areas, well depth is 30-74 ft. -- In protected bedrock areas overlain with 50 ft. of sand or gravel, well depth is 100-149 ft. -- In shallow bedrock areas, well depth is 50-199 ft. -- In Karst areas, well depth is 500-999 ft. -- OTHER **6**	CONSIDERABLE: -- Soils moderate to low in organic matter (1.5 to 0.5%). -- Moderate to rapid percolation in silty loams, loams, or silts. -- In protected bedrock areas, well depth is 15-29 ft. -- In protected bedrock areas overlain with 50 ft. of sand or gravel, well depth is 50 - 99 ft. -- In shallow bedrock areas, well depth is 25-49 ft. -- In Karst areas, well depth is 100-499 ft. -- OTHER **(4)**	HIGH: -- Soils low to very low in organic matter (less than 0.5%). -- Rapid percolation in coarse textured loamy sands or sands. -- In protected bedrock areas, well depth is less than 15 ft. -- In protected bedrock areas overlain with 50 ft. of sand or gravel, well depth is less than 50 ft. -- In shallow bedrock areas, well depth is less than 25 ft. -- In Karst areas, well depth is less than 100 ft. -- OTHER **0**	See animal waste, nutrients, pesticide, & salt "B" Field Sheets for practices

1. Add the circled Rating Item scores to get a total for the field sheet.   TOTAL [**30**]
2. Check the ranking for this site based on the total field score. Check "excellent" if the score totals at least 40. Check "good" if the score falls between 26 and 39, etc. Record your total score and rank (excellent, good, etc.) in the upper right-hand corner of the field sheet. If a Rating Item is "fair" or "poor," find the practices in the right-hand column to help remedy the conditions.

RANKING   Excellent (40-46) [   ]   Good (26-39) [**30**]   Fair (10-25) [   ]   Poor (9 or less) [   ]

Chapter 3

Ecology of Freshwater Systems

To assess properly whether or not a watercourse or water body is polluted or potentially could become polluted, you will need to know the basic ecological principles covered in this chapter.

Freshwater systems can be divided into lentic (standing) and lotic (flowing) water. Lotic systems are less prone to stress from sediment, nutrients, and pesticides because the running water flushes away pollutants. Lentic bodies, such as ponds and lakes, are more prone to pollutant stress because they retain many pollutants within their system. Impounded or dammed rivers flush out pollutants at rates which are between those for lakes and free-flowing rivers.

Lentic Systems (Lakes or Ponds)

The naturally occurring geologic process whereby lakes fill with sediment and eventually become dry land is termed "lake succession." Sediment is deposited concentrically from the outer edges to the center of the basin. Thus, a transect from the shoreline to the lake center crosses successively younger geologic sediment deposits. This concentric or horizontal zonation of sediment is reflected in concentric bands of vegetation.

Rooted aquatic plants progressively encroach toward the center from the shoreline. Large plants (macrophytes), such as cattails, alligator weed, and smartweed, generally occur in a band along the water's edge. Floating, leaved, emergent plants, such as waterlilies and American lotus root, (fig. B-7; see appendix B) occur in the bottom muds at shallow depths (0-5 feet). These plants are flanked on the inside (toward the lake/pond center) by a band of submerged rooted weeds, such as watermilfoil, coontail, and pondweed (fig. B-7). The submerged plants usually grow to a depth of about 10 feet, depending upon wave action and turbidity of the water. The region of open water is inhabited by nonrooting plants of two types, (1) microscopic floaters or plankton species (fig. B-1 to B-6), and (2) macroscopic floating species, such as duckweed (ref. 3-1, 3-2).

Associated with lake succession is eutrophication or lake enrichment by nutrients. The nutrient load of a water body is not directly observable. However, since nutrients stimulate plant growth, the biomass (total weight) of lake or pond aquatic vegetation can serve as an indirect indicator of nutrient levels. Since plants serve as food for animals, an abundance of plants often means there will be an abundance of fish and other animals. The biomass of plants and animals living in a given water body area in a unit of time is called "biological productivity."

Lentic (standing) waters are classified in biological productivity terms as: (1) "oligotrophic" (young, low productivity); (2) "mesotrophic" (middle aged, medium productivity); or (3) "eutrophic" (old, high productivity) (ref. 3-3).

Oligotrophic lakes are those which are young, geologically speaking, or are located in an infertile watershed. They are characterized by low levels of nutrients and consequently low levels of biological productivity. Having a low volume of plants (phytoplankton) contained in a large volume of water, these water bodies appear crystal clear. Since there is not much plant food at the base of the food chain, top predators, such as prized sport fish, are not abundant. Lake Superior, Lake Tahoe, and Crater Lake are examples of oligotrophic lakes. In these deep blue, clear waters fish can be seen at considerable depths from the surface.

Mesotrophic lakes are the so called "middle-aged" lakes which have a greater amount of nutrients per unit volume of water compared to oligotrophic lakes. They are more productive and have quite an abundance of organisms that are high on the food chain. For example, a 50 million pound catch of the highly edible lake trout, whitefish, blue-pike, and walleye from Lake Erie was recorded in 1920. Many of the lakes, bays and estuaries prized for their fisheries are mesotrophic (ref. 3-4, 3-5).

Eutrophic lakes have great productivity and high nutrient turnover. Water quality in these lakes with excessive nutrients can deteriorate so much that the lakes become unfit for human use. Human-induced (cultural) eutrophication may result in unsightly scums of surface algae, dead fish, and weeds washed up in mounds along the shoreline. The noxious smell of rotten eggs may result from hydrogen sulfide bubbling to the surface from the decaying organic matter.

The process of natural versus human-induced eutrophication and the presence of eutrophication indicators are discussed in more detail in Chapter 5.

Lotic Systems (Streams or Rivers)

As with plants and animals, watercourses progress through a natural life cycle from youth to old age. A young stream flows in a fairly straight path and cuts deeply into its parent soil material. In hilly terrains, it produces a narrow V-shaped valley with steep-sloped banks. As the stream matures, its path begins to meander, cutting into adjacent slopes and widening the valley. By old age, the stream has created a broad V-shaped valley and meanders back and forth within a broad flood plain (ref. 3-6).

Thus, a stream is not static, but is a delicately balanced system, ever changing in response either to natural events or to human activities. In a well-balanced "ideal" condition a stream has smooth, gentle banks—well vegetated banks free from erosion or failure—and a channel bed that is neither scouring nor building up with sediment. However, this situation seldom occurs in nature. Instead, we find streams in a continual state of adjustment, responding to the environment. It is not uncommon to find in riparian (stream bank) areas, cattle-grazing, fallen trees, or debris. Fallen trees or debris can deflect water from its main course, causing it to undercut the bank and lose vegetation. Protective vegetative cover in the watershed may be lost as land is converted to cropland or to urban development. The watercourse's adjustment to these ecological disturbances usually occurs not just at the site of the disturbance, but in domino-like fashion along a significant stretch downstream from the activities (ref. 3-7).

A watercourse adjusts to environmental effects by changing the shape of its bed, banks, or both. In an unbalanced condition, the bed will be either degrading (being scoured out) or aggrading (depositing excess sediment). Either situation is unstable and can lead to significantly adverse conditions. For example, if the bank toe is eroding, bank failure can result. If the streambed is rising, channel capacity will be reduced. In the next flood, the stream will attempt to stabilize and restore itself to its original capacity by scouring out the bed and in many cases eroding the banks as well (ref. 3-7).

Watercourse bottom materials (substrates) will vary depending upon regional geology and topography. In steep terrain, swiftly flowing waters often cut deep channels and keep the streambed scoured of sediments. By contrast, slowly flowing streams in level terrain are usually characterized by shallow beds and substrates composed mainly of sediment. Exceptions exist to the above situation, reflecting the geology of a region. For example, there are some high-velocity watercourses possessing fine bottom materials and some low-velocity watercourses with coarse bottom materials.

In general, stream flow or velocity varies according to the shape, size, slope, and roughness of the channel. Velocities range from slow (0.1 m/sec or 0.3 ft/sec); to moderate (0.25–0.5 m/sec or 0.8–1.6 ft/sec); to swift (1.0 m/sec or 3.2 ft/sec), depending on channel characteristics. Stream velocity determines in large measure the type of bottom materials present, which in turn influence the kinds and number of organisms that can live on the streambed. Erosion of sand and gravel river beds occurs at velocities greater than 1.7 m/sec (5.6 ft/sec). Gravel settles at velocities ranging from 1.2–1.7 m/sec (3.9–5.6 ft/sec). Sand settles at velocities of 0.25–1.2 m/sec (0.8–3.9 ft/sec), and silt and organics deposit when velocities drop to 0.2 m/sec (0.7 ft/sec) and less (ref 3-8).

Biology of Streams

Watercourses having cobble and gravel beds (i.e., those that are degrading or eroding) support the greatest diversity of invertebrate life. The cobble or gravel bottom is stable and provides hiding places that bottom-dwelling animals need for protection. Usually, these streams have alternating pools (deep, slow-moving water) and riffles (shallow, fast-moving water). The greatest insect production occurs in riffles with rocks of 6 in. to 12 in. on a side (ref. 3-9).

The presence of larval insect species, such as stoneflies, caddisflies, and mayflies in riffle areas of cobble/gravel bottom streams, is an indicator of "clean" water. Although the presence of these species indicates "clean" waters, absence of these species does not always mean polluted water. There are many reasons why the species might be absent. For example, they may have been exterminated by a recent flood or drought and not have had time to recolonize. Or recolonization may be impossible due to limited flight range of the insect or simply because there may be no individuals available to recolonize the location. No single insect or other invertebrate by itself can indicate pollution, but a group or association of indicator organisms can indicate the presence or absence of pollution (ref. 3-10). Refer to appendix A for biological index methods.

Aggrading or depositing streams with silt or mud bottoms support invertebrate species, such as tube-building worms, burrowing mayflies, "blood worm" midges (chironomids), mussels, and clams. The deepest parts of very large rivers, such as the Mississippi and its large tributaries, support few, if any, bottom-dwelling species because their silty bottoms are unstable.

Intermediate between cobble/gravel and mud/silt streambeds are sandy beds. Sandy bottoms support very few, if any, invertebrate species because shifting sands provide few stable surfaces to which organisms can attach.

Watercourses with slow, relatively clear waters or pools support the greatest amount of plant growth. Plants common to these waters include submerged periphyton species, such as algal or vegetative masses growing on bottom substrate materials, on twigs, or on larger rooted aquatic plants. Rooted aquatics can be either submergent species, such as *Elodea* (American waterweed), or emergents, such as the broad leaved species of *Potamogeton* (pondweed) (fig. B-7) and *Nasturtium* (watercress). These species root in the fine sediments of pools or along stream margins (ref. 3-1).

The kind and amount of aquatic vegetation in watercourses or bodies depend on a variety of factors, including flow rate, bottom type, sunlight amount, nutrient levels, and water depth. While the amounts of nutrients coming from agricultural lands might be significant, any pollutional effects from the nutrients might be minimized or "masked" by too little sunlight reaching aquatic plants for photosynthesis. Reduced sunlight can be caused by many factors, including heavy siltation of the water, dense vegetative canopy over watercourses, depth of water, etc.

Watercourses may be classified on the basis of the type of fishery they support. There are cold water, cool water, and warm water fisheries. Cold water fish include salmonid species, such as trout and salmon (fig. B-12), which are members of the trout family. These species occur in well oxygenated streams that have a swift current. Trout grow best in waters between 50 and 65 degrees Farenheit. They are insect-feeders, eating species such as mayflies and stoneflies.

The smallmouth bass (fig. B-12) is typical of cool water fisheries and is found in lower stream reaches that are marginal for trout. The bass prefer a habitat of riffles and deep pools. Home range is normally restricted to one pool where the bass feed on insects or crayfish flushed out by turtles and bottom-feeding fishes.

Where water temperatures are higher, warm water species, such as largemouth bass, crappie, bluegill, and catfish are found (fig. B-12). The largemouth bass is a predator that feeds on almost any animal which swims or falls in the water (fish, crayfish, large insects, frogs, snakes, mice). It is one of the most popular warm water fish in North America. These fish are mainly invertebrate eaters except for the catfish, which eats both plants and animals (ref. 3-11, 3-12). See appendix B for fish illustrations and descriptions.

Chapter 4

Sediment

In the United States today, watersheds are adversely affected by agriculturally related pollutants. Sediment, probably the most common and most easily recognized of the nonpoint source pollutants, ranks first in quantity among pollutants contributed by agriculture to receiving waters. Cropland erosion accounts for 40 to 50 percent of the approximately 1.5 billion tons of sediment that reaches the Nation's waterways each year. Streambank erosion accounts for another 26 percent (ref. 4-1). The amount of sediment eroding from agricultural areas is directly related to land use—the more intensive the use, the greater the erosion. For example, in a given locality more sediment erodes from row crop fields than from pastures or woodlands.

Sediment lost from agricultural sites varies significantly with the presence or absence of management practices. Figure 4-1 shows that considerably more sediment is lost from agricultural land in row crops without management practices than in row crops with management practices. The least amount of sediment is lost from agricultural lands that have conservation cropping systems, i.e., practices such as cover crops and conservation tillage (ref. 4-2).

Sediment Indicators for Receiving Waters
1. Turbidity (Refer to Field Sheet 1A, rating item 1, figure 4-2.)

To assess sediment pollution, it is necessary to observe receiving waters during or immediately following a storm event. Sediment-laden runoff, whether from overland flow or bank erosion, muddies receiving waters, and turbidity in the form of suspended solid matter increases. As turbidity increases, light penetration decreases, making objects less visible at greater depths.

If the receiving waters appear turbid, the cause must be determined. Problem sources may be overland flow paths or channels that drain from fields and pastures into receiving waters. The muddier (thicker and denser) the overland flows, the greater the sediment load. Evidence of bank erosion should be noted.

If receiving waters are turbid, but runoff water from overland flow is essentially clear (e.g., runoff from a densely vegetated pasture), and there appears to be no bank erosion

Figure 4-1

Sediment Losses Related to Land Use Practices.

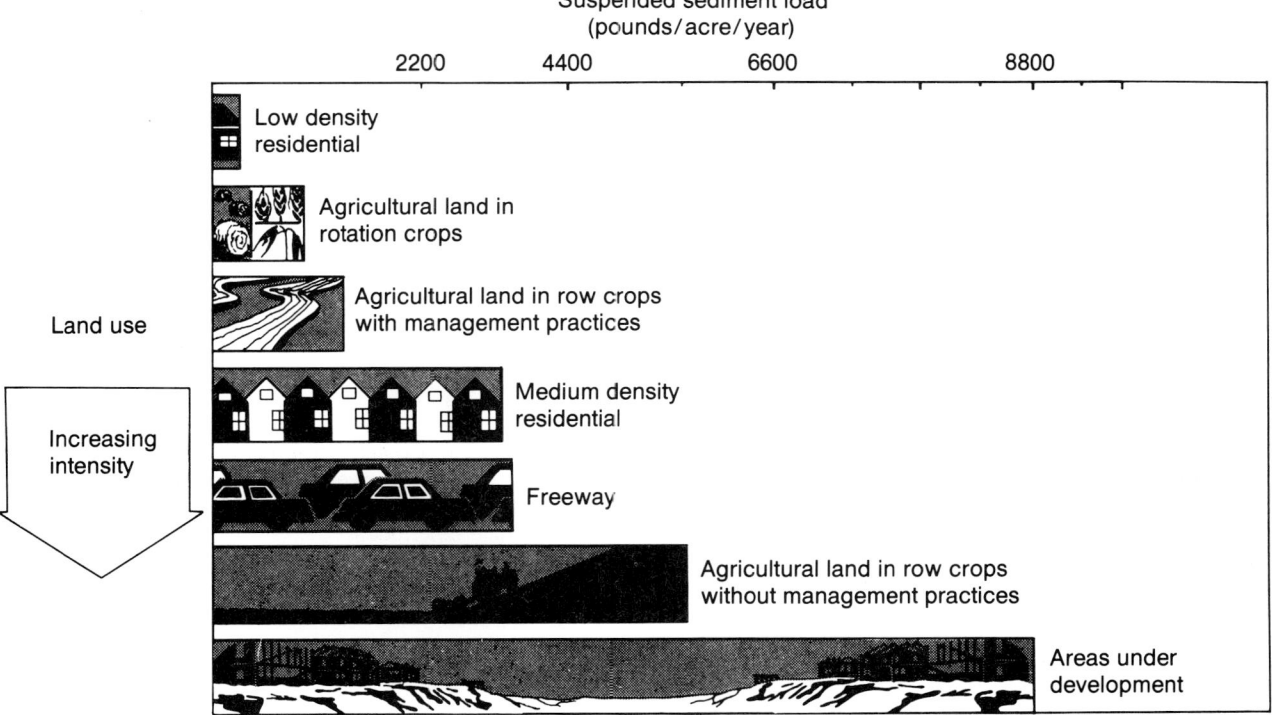

Source: Wisconsin Department of Natural Resources, Ref. 4-2.

Figure 4-2

Sediment Page 1 of 2

FIELD SHEET 1A: SEDIMENT
INDICATORS FOR RECEIVING WATERCOURSES AND WATER BODIES

Evaluator _____ County/State _____ Date _____
Water Body Evaluated _____ Water Body Location _____ Total Score/Rank _____

(Circle one number among the four choices in each row which BEST describes the conditions of the watercourse or water body being evaluated. If a condition has characteristics of two categories, you can "split" a score.)

Rating Item	Excellent	Good	Fair	Poor
1. Turbidity (best observed immediately following a storm event)	-- What is expected under pristine conditions in your region. -- Clear or very slightly muddy after storm event. -- Objects visible at depths greater than 3 to 6 ft. (depending on water color). -- OTHER **9**	-- What is expected for properly managed agricultural land in your region. -- A little muddy after storm event but clears rapidly. -- Objects visible at depths between 1½ to 3 ft. (depending on water color). -- OTHER **7**	-- A considerable increase in turbidity for your region. -- Considerable muddiness after a storm event. -- Stays slightly muddy most of the time. -- Objects visible to depths of ½ to 1½ ft. (depending on water color). -- OTHER **3**	-- A significant increase in turbidity for your region. -- Very muddy—sediment stays suspended most of the time. -- Objects visible to depths less than ½ ft. (depending on water color). -- OTHER **0**
2. Bank stability in your viewing area	-- Bank stabilized. -- No bank sloughing. -- Bank armored with vegetation, roots, brush, grass, etc. -- No exposed tree roots. -- OTHER **10**	-- Some bank instability. -- Occasional sloughing. -- Bank well-vegetated. -- Some exposed tree roots. -- OTHER **7**	-- Bank instability common. -- Sloughing common. -- Bank sparsely vegetated. -- Many exposed tree roots & some fallen trees or missing fence corners, etc. -- Channel cross-section becomes more U-shaped as opposed to V-shaped. -- OTHER **4**	-- Significant bank instability. -- Massive sloughing. -- No vegetation on bank. -- Many fallen trees, eroded culverts, downed fences, etc. -- Channel cross-section is U-shaped and stream course or gully may be meandering. -- OTHER **1**
3. Deposition (Circle a number in only A, B, C, or D)	SELECT 3A OR 3B OR 3C OR 3D			
3A. Rock or gravel streams OR	A. For rock and gravel bottom streams: -- Less than 10% burial of gravels, cobbles, and rocks. -- Pools essentially sediment free. **9**	A. For rock and gravel bottom streams: -- Between 10% & 25% burial of gravels, cobbles, & rocks. -- Pools with light dusting of sediment. **7**	A. For rock & gravel bottom streams: -- Between 25% and 50% burial of gravels, cobbles and rock. -- Pools with a heavy coating of sediment. **3**	A. For rock & gravel bottom streams: -- Greater than 50% burial of gravels, cobbles and rocks. -- Few if any deep pools present. **1**
3B. Sandy bottom streams OR	B. For sandy streambeds: -- Sand bars stable and completely vegetated. -- No mudcaps or "drapes" (coverings of fine mud). -- No mud plastering of banks; exposed parent material. -- No deltas. **9**	B. For sandy streambeds: -- Sand bars essentially stable and well, but not completely, vegetated. -- Occasional mudcaps or "drapes." -- Some mud plastering of banks. -- Beginnings of delta formation. **7**	B. For sandy streambeds: -- Sand bars unstable with sparse vegetation. -- Mudcaps or "drapes" common. -- Considerable mud plastering of banks. -- Significant delta formation. **3**	B. For sandy streambeds: -- Sand bars unstable and actively moving with no vegetation. -- Extensive mudcaps or "drapes." -- Extensive mud plastering of banks. -- Extensive deltas. **1**
3C. Mud-bottom streams OR	C. For mud bottom streams: -- Dark brown/black tanic-colored water (due to presence of lignins and tanins). -- Abundant emergent rooted aquatics or floating vegetation. **9**	C. For mud bottom streams: -- Dark brown colored water. **7**	C. For mud bottom streams: -- Medium brown water, muddy bottom. **3**	C. For mud bottom streams: -- Light brown colored, very muddy bottom. **1**

Figure 4-2

Sediment Page 2 of 2

FIELD SHEET 1A: SEDIMENT, Continued
INDICATORS FOR RECEIVING WATERCOURSES AND WATER BODIES

Rating Item	Excellent	Good	Fair	Poor
3D. Ponds	-- Ponds essentially sediment free. -- No reduction in pond storage capacity. -- OTHER 9	-- Ponds with light dusting of sediment. -- Very little loss in pond storage capacity. -- OTHER 7	-- Ponds with a heavy coating of sediment. -- Some measurable loss in pond storage capacity. -- OTHER 3	-- Ponds filled with sediment. -- Significant reduction in pool storage capacity. -- OTHER 1
4. Type and amount of aquatic vegetation & condition of periphyton (plants, growing on other plants, twigs, stones, etc.)	-- Periphyton bright green to black. Robust. -- Abundant emergent rooted aquatics or shoreline vegetation. -- In ponds, emergent rooted aquatics (e.g. cattails, arrowhead, pickerelweed, etc.) present, but in localized patches. -- OTHER 9	-- Periphyton pale green and spindly. -- Emergent rooted aquatics or shoreline vegetation common. -- In ponds, emergent rooted aquatics common, but confined to well-defined band along shore. -- OTHER 7	-- Periphyton very light colored or brownish and significantly dwarfed. -- Sparse vegetation. -- In ponds, emergent rooted aquatics abundant in wide bank; encroachment of dry land species (grasses, etc.) along shore. -- OTHER 5	-- No periphyton. -- No vegetation. -- In ponds, emergent rooted aquatics predominant with heavy encroachment of dry land species. -- OTHER 2
OPTIONAL: 5. Bottom stability of streams	-- Stable. -- Less than 5% of stream reach has evidence of scouring or silting. -- OTHER 9	-- Slight fluctuation of streambed up or down (aggradation or degradation). -- Between 5-30% of stream reach has evidence of scouring or silting. -- OTHER 7	-- Considerable fluctuation of streambed up or down (aggradation or degradation). -- Scoured or silted areas covering 30-50% of evaluated stream reach. -- Flooding more common than usual. -- More stream braiding than usual for region. -- OTHER 3	-- Significant fluctuation of streambed up or down (aggradation or degradation). -- More than 50% of stream reach affected by scouring or deposition. -- Flooding very common. -- Significantly more stream braiding than usual for region. -- OTHER 1
OPTIONAL: 6. Bottom dwelling aquatic organisms	-- Intolerant species occur: mayflies, stoneflies, caddisflies, water penny, riffle beetle and a mix of tolerants. -- High diversity. -- OTHER 9	-- A mix of tolerants: shrimp, damselflies, dragonflies, black flies. -- Intolerants rare. -- Moderate diversity. -- OTHER 7	-- Many tolerants (snails, shrimp, damselflies, dragon flies, black flies). -- Mainly tolerants and some very tolerants. -- Intolerants rare. -- Reduced diversity with occasional upsurges of tolerants, e.g. tube worms and chrionomids. -- OTHER 3	-- Only tolerants or very tolerants: midges, craneflies, horseflies, rat-tailed maggots, or none at all. -- Very reduced diversity; upsurges of very tolerants common. -- OTHER 1

1. Add the circled Rating Item scores to get a total for the field sheet. TOTAL [   ]
2. Check the ranking for this site based on the total field score. Check "excellent" if the score totals at least 32. Check "good" if the score falls between 21 and 31, etc. Record your total score and rank (excellent, good, etc.) in the upper right-hand corner of the field sheet. If a Rating Item is "fair" or "poor," complete Field Sheet 1B.

RANKING	Excellent (32-37) [ ]	Good (21-31) [ ]	Fair (9-20) [ ]	Poor (8 or less) [ ]
OPTIONAL RANKING (with #5 OR #6)	Excellent (40-46) [ ]	Good (26-39) [ ]	Fair (11-25) [ ]	Poor (10 or less) [ ]
OPTIONAL RANKING (with #5 AND #6)	Excellent (48-55) [ ]	Good (31-47) [ ]	Fair (13-30) [ ]	Poor (12 or less) [ ]

(e.g., banks well vegetated), the turbidity may be due to stirred-up mud deposits of the stream bottom. This is common in regions characterized by muddy-bottom streams. In this situation, the regional environmental quality would be considered "excellent" despite the muddiness because conditions match what is expected under pristine conditions in that particular geographic region.

2. **Bank stability** (Refer to Field Sheet 1A, rating item 2, figure 4–2.)

   To determine if streambanks are contributing sediment to receiving waters, look for the following indicators:

   - Evidence of bank instability—cracks, rills, and gullies.
   - Evidence of bank sloughing or chunks of soil dropping into the stream.
   - Extent of vegetative protective cover or "armoring."
   - Extent of exposed tree roots, fallen trees, missing fence posts, etc.
   - The appearance of the channel in cross-section (adapted from Keown, ref. 3-7).

3. **Deposition** (Refer to Field Sheet 1A, rating item 3, figure 4–2.)

   Watercourses are distinguished on the basis of their type of bottom substrate—rock, gravel, sand, or mud. Deposition occurs when water flow is insufficient to remove sediment entering receiving waters.

   Note that this field sheet gives four choices for deposition. Items 3A, 3B, and 3C refer to streams or flowing waters, while item 3D refers to ponds or stationary waters.

   Indicators of deposition vary with the type of bottom substrate. In rock and gravel streams, the relative degrees of burial of gravels, cobbles, and rocks in riffle (fast-flowing, shallow) areas are important as well as the thickness of sediment coatings in pool areas (see item 3A). For sandy streambanks the condition and stability of sandbars and the presence and frequency of mudcaps (drapes), mud plastering, and delta formation are important (see rating item 3B). For mud-bottom streams, water color is especially important (rating item 3C). In this case, it is essential to be familiar with waters of your region. You can gain familiarity with the "normal" color of local streams quickly by several onsite visits before and immediately following a storm event. Finally, indicators of pond degradation are thickness of the sediment coat and the relative degree of reduction in permanent pond storage capacity (see rating item 3D).

4. **Type and amount of aquatic vegetation and condition of periphyton** (plants growing on other plants, twigs, stones, etc.) (Refer to Field Sheet 1A, rating item 4, figure 4–2).

   In those waters where aquatic vegetation is typical of that expected under pristine conditions in your geographic region, sediment load may become great enough to interfere with plant growth and reproduction. For example, periphyton (small aquatic plants that grow on submerged plants, twigs, stones, etc.) may create a "dusting" or coating on aquatic plants, reducing their photosynthesis. Sediment (silt) may also accumulate on aquatic plants and add to the poor environmental conditions. Aquatic plants may appear to be paler green and more spindly than the robust green condition that is found where light penetration is maximal. Where there is considerable sediment deposition, aquatic plants may never reach full size and are not able to reproduce. Eventually, as occurred in the Chesapeake Bay, an entire population of aquatic plants may smother and die.

5. **Bottom stability of watercourses** (Refer to Field Sheet 1A, rating item 5, figure 4–2).

   In instances where historical records are available, bottom stability might serve as an indicator of sediment pollution. For example, aggradation (raising of streambeds) is an indicator of sediment deposition. Deposition is sometimes greatly accelerated by logjams or other stream obstructions. These obstructions can slow water to an extent that sediment that usually is flushed through the system has time to settle out. Given enough time, this type of deposition can lead to a significant rise in the streambed with a number of attending consequences. One consequence is that the flow becomes shallow and spreads out over a wide area, resulting in increased flooding and increased stream "braiding," the formation of many small rivulets. It may also result in the death of economically valuable bottomland hardwood trees. In such instances, it may be necessary to dredge or dynamite a channel to restore water flow to its original depth. An increased need for dredging is a good indicator that sediment deposition has increased (ref. 4–3).

Chapter 5

Nutrients

Natural and Human-Induced (Cultural) Eutrophication

Eutrophication is a natural aging process that occurs as a lake or pond becomes increasingly enriched with nutrients. The rate of eutrophication varies, depending upon the relative fertility of the watershed. It proceeds most slowly in big lakes situated in relatively infertile watersheds and most quickly in small ponds in fertile surroundings.

Eutrophication can be natural or human-induced (fig. 5-1). Eutrophication, resulting from human activity, such as fertilizing fields or converting forest or pasture to cropland, is termed human-induced (cultural) to distinguish it from natural eutrophication. In most instances, the rate of human-induced eutrophication is many times faster than the natural process. For example, in a span of about 25 years (1950-1975), Lake Erie aged to about the same degree under human influences as would have occurred in 15,000 years naturally (ref. 5-1). Today, some 75 percent of the large lakes in the United States are considered to be eutrophic (ref. 5-2).

Eutrophication rates are increased by agricultural inputs of nutrients—phosphorus and/or nitrogen. Usually, these inputs come from either fertilizer runoff or erosion from fields or pastures.

Indicators of Excessive Nutrient Input for Receiving Waters

LIMITATIONS OF NUTRIENT FIELD SHEET 3A

Nutrient indicators may not be perceptible in certain watercourses, especially if flow is 0.5 feet per second or greater or if sediment "masks" the effects of nutrient enrichment. Appendix A contains a procedure ("Floating Body Technique") that can be used to obtain water flow rate velocity. With rapid water flow, a watercourse could be rated "good" or "excellent" according to the 3A Nutrient Field Sheet (fig. 5-2), when in fact it could contain high nutrient levels.

In the above situations it may be advantageous to use the 3B Nutrient Field Sheet first to determine if present agricultural management practices may be contributing to nutrient enrichment in the nearby watercourse.

Additionally, it may be necessary to conduct or have conducted nutrient chemical analyses or to contact the State water quality agency to get nutrient values for the watercourse being examined.

Figure 5-1

Advanced Eutrophication of a Pond.

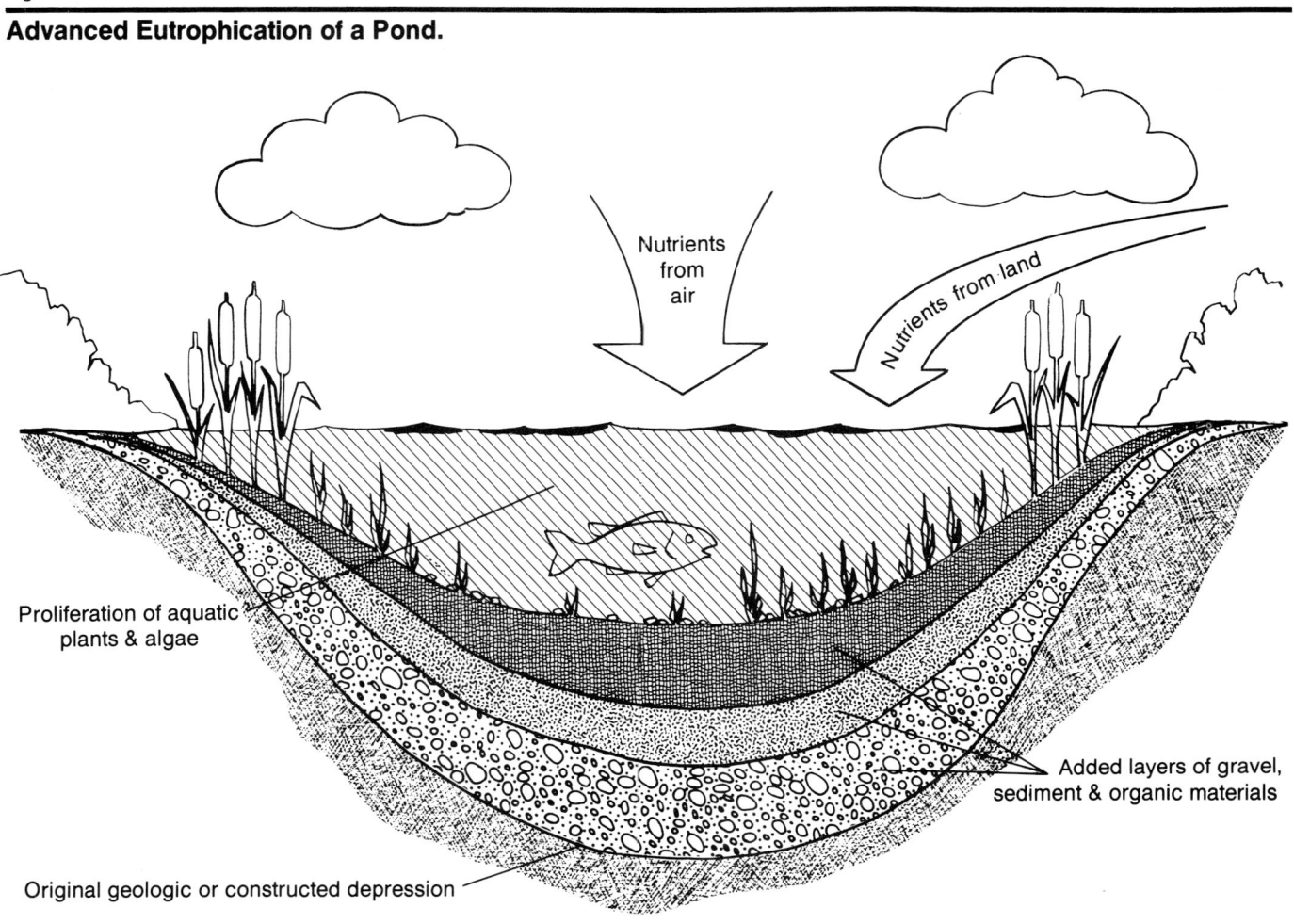

Figure 5-2

Nutrients

FIELD SHEET 3A: NUTRIENTS
INDICATORS FOR RECEIVING WATERCOURSES AND WATER BODIES*

Evaluator _____ County/State _____ Date _____
Water Body Evaluated _____ Water Body Location _____ Total Score/Rank _____

(Circle one number among the four choices in each row which BEST describes the conditions of the watercourse or water body being evaluated. If a condition has characteristics of two categories, you can "split" a score.)

Rating Item	Excellent	Good	Fair	Poor
1. Total amount of aquatic vegetation at low flow or in pooled areas. Includes rooted and floating plants, algae, mosses & periphyton	-- Little vegetation, uncluttered look to stream or pond. OR -- What's expected for good water quality conditions in your region. -- Usually fairly low amounts of many different kinds of plants. -- OTHER __ 10	-- Moderate amounts of vegetation. OR -- What's expected for good water quality conditions in your region. -- OTHER __ 6	-- Cluttered weedy conditions. Vegetation sometimes luxurious and green. -- Seasonal algal blooms. -- OTHER __ 3	-- Choked weedy conditions or heavy algal blooms or no vegetation at all. -- Dense masses of slimy white, greyish green, rusty brown or black water molds common on bottom. -- OTHER __ 0
2. Color of water due to plants at base or low flow	-- Clear or slightly greenish water in pond or along the whole reach of stream. -- OTHER __ 9	-- Fairly clear; slightly greenish. -- OTHER __ 6	-- Greenish. Difficult to get pond sample without pieces of algae or weeds in it. -- OTHER __ 3	-- Very, very green pond scums. -- Pea green color or pea soup condition during seasonal blooms of microscopic algae in ponds. -- "Oily-like" sheen when pea soup algae die off. -- OTHER __ 0
3. Fish behavior in hot weather fish kills, especially before dawn	-- No fish piping or aberrant behavior. -- No fish kills. -- OTHER __ 9	-- In hot climates, occasional fish piping or gulping for air in ponds just before dawn. -- No fish kills in last two years. -- OTHER __ 5	-- Fish piping common just before dawn. -- Occasional fish kills. -- OTHER __ 3	-- Pronounced fish piping. -- Pond fish kills common. -- Frequent stream fish kills during spring thaw. -- Very tolerant species (e.g. bullhead, catfish). -- OTHER __ 0
4. Water use impacts; health effects for whole sub-watershed	-- None. -- OTHER __ 8	-- Minimal, such as reduced quality of fishing. -- OTHER __ 7	A couple of the following: -- Algal clogged pipes. -- Algal related taste, color, or odor problems with human or livestock water supply. -- Cattle abortion. -- Reduced recreational use due to weedy conditions, decay, odors, etc. -- OTHER __ 4	Several of the following: -- Algal clogged pipes. -- Algal related taste, color, or odor problems with human or livestock water supply. -- Cattle abortion. -- Reduced quality of fishery. -- Reduced recreational use due to weedy conditions, decay, odors, etc. -- Blue babies—incidence of methemoglobinemia due to high nitrate levels. -- Property devaluation. -- OTHER __ 2
5. Bottom-dwelling aquatic organisms	-- Intolerant species occur: mayflies, stonefiles, caddisflies, water penny, riffle beetle. -- High diversity. -- OTHER __ 9	-- Intolerants common. -- A mix of tolerants: shrimp, damselflies, dragonflies, black flies. -- Moderate diversity. -- OTHER __ 7	-- Mainly tolerants: snails, shrimp, damselflies, dragonflies, black flies. -- Mainly tolerants, but some very-tolerants. -- Intolerants rare. -- Reduced diversity with occasional upsurges of tolerants, e.g. tube worms, and chironomids. -- OTHER __ 3	-- Mainly very-tolerants: midges, craneflies, horseflies, rat-tailed maggots, or no organisms at all. -- Very reduced diversity, upsurges of very-tolerants common. -- OTHER __ 1

*The effects of nutrients may be "masked" by high sediment loads, creating sufficient turbidity to shade light-dependent aquatic vegetation. This may cause aquatic vegetation, a water quality indicator, to die and disappear from the watercourse. To obtain accurate nutrient levels in high sediment situations, chemical testing may be necessary. Under these circumstances you should contact a local or other water quality specialist.

1. Add the circled Rating Item scores to get a total for the field sheet. TOTAL [   ]
2. Check the ranking for this site based on the total field score. Check "excellent" if the score totals at least 38. Check "good" if the score falls between 23 and 37, etc. Record your total score and rank (excellent, good, etc.) in the upper right-hand corner of the field sheet. If a Rating Item is "fair" or "poor," complete Field Sheet 3B.

RANKING   Excellent (38-45) [   ]   Good (23-37) [   ]   Fair (9-22) [   ]   Poor (8 or less) [   ]

1. **Total amount of aquatic vegetation** (Refer to Field Sheet 3A, rating item 1, figure 5-2.)

   Aquatic vegetation must be supplied with a sufficient quantity of nutrients to grow and reproduce. Vegetative growth in many waterways and bodies is held in check by a limited amount of an available nutrient, i.e., the limiting nutrient. Typically, waters are phosphorus limited, although in some areas the waters naturally contain high phosphate levels and nitrogen is the limiting nutrient.

   Agriculturally related inputs of phosphorus, nitrogen, or both to nutrient-limited waters promote aquatic plant growth. With minimal additions of nutrients, plants may appear even more robust and luxurious than usual. For example, watercress that has additional nutrients may be darker green than normal. By contrast, moderate amounts of nutrients may result in noticeable increases in plant biomass. Stands of watercress under this condition might enlarge considerably in surface area. Heavy additions of nutrients can stimulate weedy proliferations or extensive algal blooms. Sometimes this potential is not realized, such as when sediment loads are so great that light becomes the limiting factor for plant growth. In this instance, sediment masks the expected effects of nutrient enrichment.

   When a watercourse or water body regularly displays symptoms of heavy nutrient enrichment, such as extensive algal slimes (scums) or weedy proliferations, it is labelled "eutrophic." It is common for these eutrophic waters to be clogged with vegetation. In general, standing bodies of water are more prone to eutrophication than flowing waters, although even streams may appear quite clogged during periods of low flow.

   Many types of aquatic vegetation, such as watermilfoil and many algae, die back at the end of summer in response to unidentified seasonal environmental influences. When significant masses of vegetation die simultaneously, the biochemical oxygen demand (BOD) of the water increases dramatically and the amount of dissolved oxygen (DO) drops precipitously as oxygen-requiring (aerobic) micro-organisms begin the process of decomposition. These lowered DO levels stress all aquatic organisms, both animals and plants, and may lead to fish kills and the elimination of all vegetation. This is discussed further in the next section.

2. **Color of water** (Refer to Field Sheet 3A, rating item 2, figure 5-2).

   Excessive growth of microscopic plants or algae (phytoplankton, figs. B-1 to B-6) often manifests itself as a change in the color of the water. Ponds in particular might assume a deeper color of various shades of green, blue-green, red, gray, or yellow depending upon the phytoplankton species present. Blue-green algae can undergo tremendous growth in numbers when phosphorus is added, so that the water can become like pea soup. Furthermore, blue-greens can survive nitrogen deficient conditions because they are able to utilize atmospheric nitrogen in much the same manner as soil bacteria in the nodules of legumes. In addition, many blue-greens secrete toxins or foul-tasting chemicals, making them most unattractive as food to other organisms.

   Animal plankton (zooplankton— small, floating or feebly swimming animals), such as water fleas, rotifers, and copepods, which usually graze on the phytoplankton (plant plankton) avoid blue-green algae. As a result, blue-green algae can grow unchecked by predators until the algae die in massive amounts. The decay of algal overgrowths leads to fluctuating oxygen levels and to periodic oxygen depletions (anoxia) that sometimes result in fish kills (fig. 5-3). During extended periods of anoxia, vegetation of all types is destroyed during the nights, when photosynthesis does not occur (ref. 5-3).

3. **Fish diversity, behavior and fish kills** (Refer to Field Sheet 3A, rating item 3, figure 5-2.)

   Nutrient enrichment can lead to the simplification of food webs by the elimination of sensitive species, which are the least able to cope with adverse conditions. Long-lived organisms that reproduce slowly and require extended periods of stable conditions fare worst in unstable eutrophied waters. In particular, fish populations often shift from dominance by larger, top predator game species to dominance by smaller, less desirable forage (rough) species. For example, in Lake Erie the long-lived highly edible sport fish, such as lake trout, whitefish, pike, and walleye, were replaced by "rough" fish—carp, smelt, drum, and alewives (fig. B-12, ref. 5-1.)

   Sensitive species, such as sport fish, decline because they cannot tolerate the periodic episodes of:

   - Low dissolved oxygen levels (anoxia) due to the decomposition by micro-organisms of massive amounts of dead plants;

   - Toxicity due to the release of the poisonous gases (hydrogen sulfide and methane) by anaerobic micro-organisms during anoxic conditions;

   - Toxicity due to secretions from some blue-green or dinoflagellate algal blooms; or

   - Some combination of the above activities with other major agricultural pollutants (adapted from Luoma, ref. 5-3).

   The loss of species diversity, as sensitive species die, is undesirable for both economic and ecological reasons. The loss of sport fish from a lake may constitute a major economic loss to sport fishermen and local businesses dependent upon the fishermen.

   Ecologically, simplification of a food web is a "warning signal" or indicator that the whole ecosystem is unhealthy and may be in jeopardy. An unhealthy system is more vulnerable than a healthy "diverse" system to further disruptions or disturbances, whether natural or caused by human activities.

Figure 5-3

The Eutrophication Process.

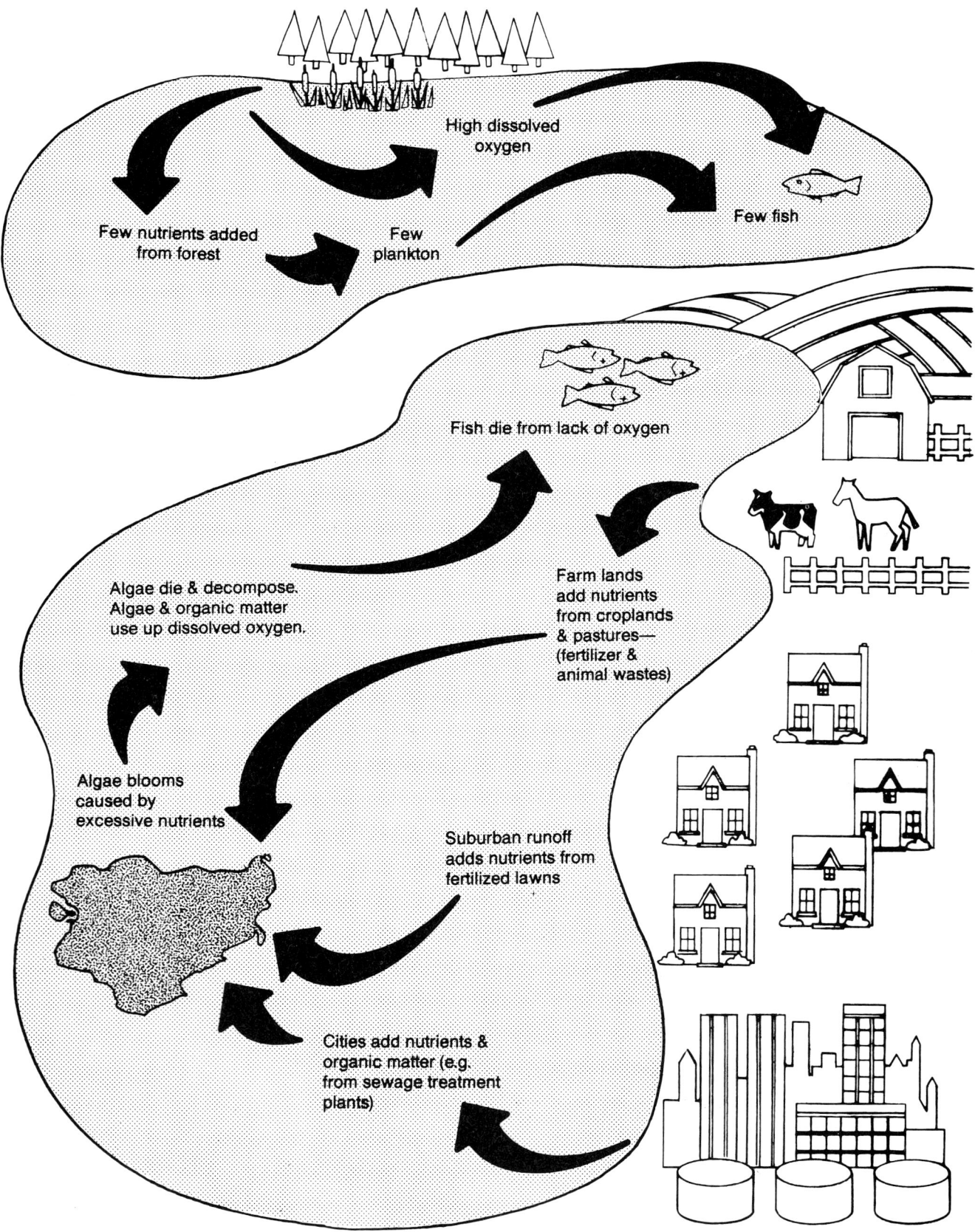

26

Fish kills can occur in ponds that receive excessive nutrient inflows. Three common scenarios for eutrophied ponds are described below, namely:

- Floating Plant (macrophyte) Infestation
- Algal Mats (filamentous) Infestation
- Pea Soup (phytoplankton) Infestation

Floating plant (macrophyte) infestation. In the summer months, floating plants, such as duckweed (*Lemna*, fig. B–7), can proliferate in ponds enriched by the runoff from fertilized fields or pastures. If left unchecked, these plants can multiply and cover the entire pond surface. When this happens, light cannot penetrate through the surface plant cover. Without light, the naturally occurring phytoplankton (microscopic algae) at the base of aquatic food chains cannot carry on photosynthesis, and little or no oxygen is produced. The protective cover of floating plants also reduces wave action, an important source of oxygen. Oxygen is depleted by the respiration of plants, animals, and micro-organisms.

Hot weather intensifies the problem by accelerating both the rate of respiration of the organisms and the chemical oxidation of many substances. Eventually, fish and other oxygen-requiring (aerobic) organisms suffocate from a lack of dissolved oxygen, and fish kills occur.

Algal mat (filamentous) infestation—fish piping common. Many farmers routinely treat ponds for floating plants before the plant populations reach nuisance proportions. However, some ponds that appear "clear" (you can see to the bottom) will have significant amounts of filamentous algae (pond scum) growing along the bottom and sides or attached to rocks or other larger plants (fig. B–7). In response to an unidentified trigger, these filamentous algae rise to the surface in mats and die, creating decaying odors and nuisances.

The sudden existence of such large quantities of dead algae in a pond pollutes the pond by increasing oxygen-demanding organic matter, which increases the biochemical oxygen demand (BOD) of the decay micro-organisms. This results in an immediate drain on the dissolved oxygen (DO). DO levels in the pond become critical at night when photosynthesis by any remaining living plants comes to a halt. The lowest DO levels occur at dawn.

At sunrise, fish in a pond with insufficient DO can be observed congregating at the edge of the water where DO levels are highest. The fish usually assume a hanging position at approximately a 45 degree angle and pipe (suck or gulp) air. Under these critical DO concentrations, fish begin to die slowly. It takes about a week of nightly DO levels dropping to levels of less than 2.0 parts per million (ppm) to achieve a total kill.

Under highly enriched conditions, aerobic decay micro-organisms may become too overworked to handle the increased organic load and may die of suffocation when DO levels approach zero. The decomposition process is then taken over by much less efficient anaerobic bacteria that do not require oxygen. These bacteria release a gas that smells like rotten eggs (hydrogen sulfide), as well as other poisonous breakdown products. The bacteria contribute to the ultimate decline of a lake or pond, which then is most unappealing in terms of sight, smell, and taste. This situation can be particularly dangerous in lakes or ponds used as reservoirs for drinking water.

Pea soup (phytoplankton) infestation. Farm ponds, which become highly enriched with nutrients may undergo much photosynthesis and take on a pea soup appearance to a depth of more than 1 ft. During summer, in some farm ponds in the South, SCS personnel have recorded supersaturated DO levels ranging up to 28 ppm at 4 o'clock in the afternoon, dropping to near 0 ppm by an hour after sunrise of the following day. Fish kills are common under such conditions.

The organisms responsible for the fish kills in the pea soup condition are phytoplankton (small, floating plants), which are so small that they can be observed only with a microscope. The phytoplankton consist of a variety of algae, including diatoms and green and blue-green algae (cyanobacteria, fig. B–1 to B–6). Despite their small size, populations of these plankton can reach proportions that color the water pea green and thicken it to resemble soup.

4. **Water use impacts** (Refer to Field Sheet 3A, rating item 4, figure 5–2).

    Agriculturally related nutrient enrichment and eutrophication can adversely affect a number of water uses. For example, eutrophied water can alter the color, taste, and odor of a drinking water supply. The removal of excessive algal slimes may also increase the cost of water treatment. Nuisance levels of vegetation or algae may detract from the aesthetic quality of the water, clog pipes and intakes, and reduce property values and recreational use.

    Finally, high nitrate levels in drinking water are known to affect adversely the health of babies and the elderly. Babies who receive too much nitrate from the water used in preparing formula may suffer from methemoglobinemia, or blue baby syndrome. Some babies have died from this condition, when it was not treated in time. These same conditions can affect the young of cattle.

5. **Bottom-dwelling aquatic organisms** (Refer to Field Sheet 3A, rating item 5, figure 5–2).

    As waters become increasingly eutrophied, the abundance and species composition of bottom organisms change. Waters receiving few, if any, excess nutrients from agricultural or other sources are characterized by a high diversity of bottom-dwelling organisms. Generally, in these very pristine waters, the diversity of bottom species is high, but the number of each type is low.

    Among bottom organisms found to be sensitive to or intolerant of nutrient excesses are mayflies, stoneflies, caddisflies, water-penny, and riffle beetles (ref. 5–4). Generally, as nutrient quantities increase, populations of these intolerant species recede. They are replaced by expanding populations of nutrient-tolerant species, such as chironomids and blackflies. The usual pattern is that as nutrients increase over time, the number of species (species diversity or richness) decreases, while the population growth of a few species increases.

An excellent tool for determining the diversity of bottom-dwelling invertebrates is the Sequential Comparison Index (SCI, appendix A), which is designed for nonprofessionals and assumes no background knowledge of biology or taxonomy (ref. 5-5). Appendix A also contains Beck's Biotic Index procedure, which requires the identification of pollution-tolerant and intolerant species to make a water quality determination. Appendix B contains pictorial keys for common invertebrates and another procedure entitled "Simple Assessment of Bottom-Dwelling Insects."

Chapter 6

Pesticides

Most agricultural pesticides are either herbicides, which make up approximately one-half of the U.S. pesticide usage, or insecticides, which make up about one-third of the pesticide usage.

Effects of Pesticides on the Aquatic Environment

The effect of a pesticide on the aquatic environment depends upon many factors, including the physical, chemical, and biological properties of the pesticide; the amount, method, and timing of application; and the intensity of the first storm event following application. In general, pesticide effects on aquatic organisms vary with the relative toxicity of the pesticide, its persistence or how long it remains active in the environment, and its tendency to accumulate in the food chain. The longer a pesticide persists in the soil, the greater the opportunity for it to be transported from the crop area to receiving waters or to ground water, or for it to affect nontarget organisms, such as animals, humans, and noncrop plants adversely.

Insecticides: Chlorinated hydrocarbon insecticides, such as DDT, which appeared after World War II, are of low-to-moderate toxicity and are termed "wide-spectrum" (i.e., they kill a wide variety of insects). These insecticides severely affected many environments, killing top-of-the-food-chain predator birds, such as the bald eagle, brown pelican, and peregrine falcon.

The most infamous of the synthetic organics was DDT. DDT is very persistent, with a half-life in sediments of greater than 10 years. The half-life is how long it takes for half of a compound to decay. Since DDT is fat soluble, it concentrates in the fat of organisms. Figure 6-1 illustrates the increase in concentration of DDD, a close relative of DDT, as DDD is passed from one organism to another up the food chain in a lake.

In some ecosystems, DDT can become concentrated at the top of the aquatic food chain in quantities great enough to interfere with reproduction or cause death. Consequently, decline or death of birds of prey at the top of aquatic food chains (e.g., bald eagle or fish hawk) may be an indicator of pesticide damage to an aquatic ecosystem.

The decline or death of sensitive fish species and other aquatic organisms also serves as an indicator of pesticide pollution. Salmonids (rainbow trout, brown trout, and salmon) are the most sensitive to chlorinated hydrocarbon pesticides. Redear sunfish, bluegill, and largemouth bass are intermediate in sensitivity, with channel catfish and black bullheads being the least sensitive (ref. 6-1).

Today, most synthetic organic insecticides have been replaced by the organophosphate insecticides (e.g., malathion and diazinon) and by carbamates (e.g., carbaryl). Organophosphate insecticides are much less persistent, with half-lives from 1 to 12 weeks. The main feature of organophosphate insecticides is their rapid degradability (ref. 6-1). Some carbamates persist only a few days.

Since carbamates and organophosphates are not fat soluble, they do not concentrate in organisms nor do they accumulate up the food chains. Consequently, the compounds are much safer environmentally. However, while organophosphate compounds are safer environmentally, a toxic organophosphate compound can kill fish in a water body and quickly degrade with no detectable chemical trace a few weeks later. Fish families still show the same types of sensitivity to the organophosphates that they did to the chlorinated hydrocarbons, with salmonids being

Figure 6-1

Biomagnification of DDD in the Food Chain at Clear Lake, California.

Numbers are times amount in water.

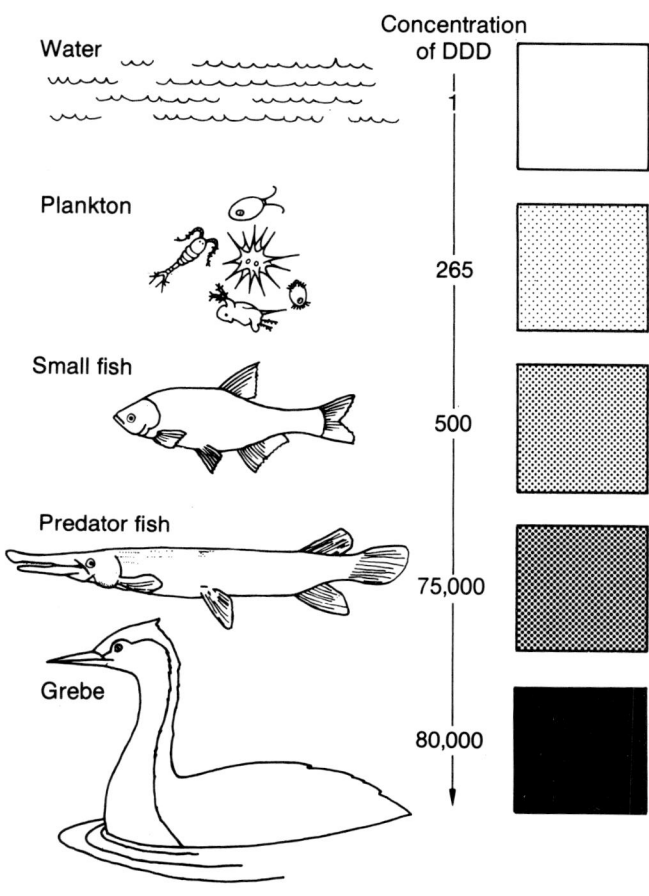

(Flint and van den Bosch, 1977) (Ref. 6-10).

the most sensitive and catfish the least sensitive. Carbamates and organophosphates are soluble in water and can be easily transported in water. Thus, these compounds may increase the potential for ground water contamination.

Herbicides: Herbicides vary considerably in their persistence. Herbicides, such as 2,4-D and alachlor, are considered to be nonpersistent, with half-lives of less than 20 days. They seldom remain in the soil for longer than a month to 6 weeks when used at the recommended levels for weed control. Atrazine is considerably more persistent, remaining in the soil for as long as 17 months. Others such as monuron, picloram, simazine, and paraquat are very persistent, remaining in the soil from 2 to 4 years. Most herbicides are nonpersistent, breaking down by the end of the growing season (ref. 6-2, 6-3).

In general, when compared to insecticides, herbicides in use today rank lower in relative fish toxicity and the potential for environmental impact. Many herbicides do not appear to have a permanent impact on aquatic ecosystems and appear to be only moderately toxic to fish.

Pesticide Indicators for Receiving Waters

1. **Presence of pesticide containers** (Refer to Field Sheet 4A, rating item 1, figure 6-2).

    Evidence of careless and haphazard disposal or dumping of pesticide containers in or near sink holes, streams, or water bodies should be a warning of possible adverse pesticide effects on the aquatic ecosystem.

2. **Appearance of nontarget vegetation** (Refer to Field Sheet 4A, rating item 2, figure 6-2).

    By definition, herbicides are toxic to plant life. Herbicides draining from agricultural fields can result in the death of aquatic vegetation. This is especially true if a storm occurs immediately following spraying and washes the newly applied pesticide into nearby waters. Also, aerial drift that carries pesticide away from the field, and "overspray" by the spray plane beyond the field edge can damage or kill aquatic vegetation by landing directly on it. Large (macroscopic) aquatic plants are particularly sensitive. Microscopic phytoplankiton appear to be less sensitive, although the effects on plankton have not been extensively studied (ref. 6-3).

    Leaf burn and evidence of vegetative dieback on nontarget vegetation, whether aquatic or terrestrial, are indicators of herbicide damage. Care should be taken to look for this type of evidence in or along ponds, drainage ditches, and streams. Examine floating species, such as pond lilies and duckweed. Also examine emergent rooted aquatics, such as watercress, watershield, and spatterdock, and marginal weeds, such as alligator weed, smartweed, arrowhead, pickerelweed, and cattails (fig. B-7).

3. **Overall diversity of insects, presence of "fish bait types"** (Refer to Field Sheet 4A, rating item 3, figure 6-2).

    Insecticides kill nontarget, as well as target insects and other closely related species. It is not uncommon to observe reduced species diversity and reduced populations of aquatic bottom-dwelling organisms in waters that receive pesticide runoff. Diversity is reduced as sensitive species, such as some types of mayflies, dragonflies, water mites, or beetles, decline or die off. As sensitive species die, populations of insecticide-tolerant species, such as blackflies and chironomids, expand to fill the void vacated by the sensitive species. Ask the landowner if there have been any insect population upsurges or decreases in the local area. An excellent tool for determining the diversity of bottom-dwelling invertebrates is the Sequential Comparison Index (SCI) shown in appendix A.

4. **Overall diversity of fish** (Refer to Field Sheet 4A, rating item 4, figure 6-2).

    Chronic sublethal effects of pesticides in waters are difficult to observe. Chronic effects include:

    - Fish avoidance of contaminated watercourse areas. This may result in their absence in a localized area or prevent their swimming into these areas to spawn.

    - Altered reproduction due to toxicity or avoidance. Trout do not naturally reproduce in some agriculturally drained streams common to their range.

    - Lowered fish productivity.

    - Young fish mortality (decreased survival of newly hatched fish; adapted from Pimentel, Brown, Cross, ref. 6-4, 6-5, 6-6).

    These subtle effects, combined with the dieback of fish food (fish bait) organisms as described in number 3 above, manifest themselves in altered or degraded fisheries. In general, the greater the input of pesticides, the less diverse the fishery. Salmonids appear to be most sensitive and decline or are eliminated first. Next in sensitivity are the intolerant centrachids, such as longear sunfish, striped bass, smallmouth bass, crappie, redfin pickerel, and bluegill. These are followed by the more tolerant centrachids (blacknose dace, common shiner, sculpin, creek chub, madtom, golden shiner, largemouth bass, blueback herring and alewives). In the worst of the chronically polluted pesticide waters, there are only the very most tolerant species of cyprinid minnows and ictalurids. Typical species include brownhead carp, bullheads, white sucker, shad and catfish, or no fish at all. See appendix B for a brief summary of fish species (ref. 6-7, 6-8).

5. **Fish kills, animal teratology** (Refer to Field Sheet 4A, rating item 5, figure 6-2).

    Acute effects of lethal concentrations of pesticides result in insect kills (mayfly, dragonfly, etc.) or fish kills or both. These kills are usually of a limited nature and are easy to observe. They frequently occur after routine spraying of barns or feedlot areas that are in close proximity to a watercourse or pond, or from the improper washing of spray equipment and containers. Massive kills are rare.

    Chronic sublethal concentrations of pesticides are sometimes teratogenic; that is, they produce birth defects or tumors. One type of birth defect that might be observed is broken-back syndrome in fish (vertebral deformities and scoliosis). Other types are deformed bird beaks or the absence of ears or eyes, resulting from elevated levels of selenium or other trace elements or toxic ions.

Figure 6-2

Pesticides

FIELD SHEET 4A: PESTICIDES
INDICATORS FOR RECEIVING WATERCOURSES AND WATER BODIES

Evaluator _____ County/State _____ Date _____
Water Body Evaluated _____ Water Body Location _____ Total Score/Rank _____

Rating Item	Excellent	Good	Fair	Poor

(Circle one number among the four choices in each row which BEST describes the conditions of the watercourse or water body being evaluated. If a condition has characteristics of two categories, you can "split" a score.)

Rating Item	Excellent	Good	Fair	Poor
1. Presence of pesticide containers	-- No containers in or near water. -- OTHER **9**	-- No containers in or near water. -- OTHER **9**	-- Containers located near the water. -- OTHER **5**	-- Containers in the water. -- OTHER **3**
2. Appearance of non-target vegetation	-- No leaf burn. -- No vegetation dieback. -- OTHER **9**	-- Some leaf burn. -- No vegetation dieback. -- OTHER **6**	-- Significant leaf burn. -- Some vegetation dieback. -- OTHER **4**	-- Severe dieback of vegetation. -- OTHER **1**
3. Overall diversity of insects ("fish bait")	-- High diversity including dragonflies, stoneflies, mayflies, caddisflies, water mites or beetles. -- OTHER **10**	-- Average diversity of insects—some of those listed under excellent. -- OTHER **8**	-- Occasional insect kills. Reduced numbers and kinds. Upsurges of blackflies & chironomids. -- OTHER **3**	-- Insect kills common. Not many fish-bait types such as hellgrammites (the larvae of dobsonflies), alderflies, or fishflies. -- OTHER **1**
4. Overall diversity of fish	-- Excellent fish diversity—what's expected in the area. -- Presence of intolerants such as brook, brown or rainbow trout, salmon or stickleback. -- OTHER **9**	-- Good fish diversity. -- Native salmonids (trout & salmon) begin to die out first. The least tolerant centrarchids (longear sunfish, rock bass, smallmouth bass, crappie, redfinned pickerel and bluegill) begin to decline. -- OTHER **7**	-- Reduced fish diversity. -- The more tolerant centrarchids die off—blacknose dace, common shiner, sculpin, creekchub, madtom, golden shiner, large mouth bass, blueback herring, and alewives. -- Larger proportion of green sunfish. -- Occasional (once every 1-2 years) pond fish kills. -- OTHER **4**	-- Extremely reduced fish diversity. -- Only very tolerant species of cyprinids & ictalurids (such as brownhead carp, bullheads, white sucker, shad, and catfish. -- Some highly polluted waters (usually ponds) may lack fish entirely. -- OTHER **1**
5. Fish kills; animal teratology (birth defects & tumors in fish & other animals)	-- No fish kills in last 2 years. -- No birth defects of tumors. -- OTHER **9**	-- Fish kills rare in last 2 years. -- Minimal birth defects & tumors occurring in the population randomly. -- OTHER **5**	-- Occasional fish kills. -- Some birth defects & tumors. -- OTHER **3**	-- Fish kills common in last couple of years. -- Frequent fish kills during spring thaws. -- Yearly pond fish kills following aquatic vegetation dieback not uncommon. -- Considerable numbers of birth defects & tumors. -- OTHER **0**
OPTIONAL 6. Fish behavior in hot weather; fish kills, especially before dawn	-- Normal behavior, e.g. fish seen breaking the surface for insects. -- No evidence of disease, tumors, fin damage, or other anomalies. -- No fish piping or aberrant behavior. -- No fish kills. -- OTHER **9**	-- Behavior as expected, e.g. evidence of fish, such as water rings or bubbles. -- Little if any evidence of disease, tumors, fin damage, or other anomalies. -- In hot climates, occasional fish piping or gulping for air in ponds just before dawn. -- No fish kills in last 2 years. -- OTHER **7**	-- Behavioral changes in fish—swimming near surface, uncoordinated movements, convulsive darting movements, erratic swimming up & down or in small circles, hyperexcitability (jumping out), difficulty in respiration. More likely seen in ponds. -- Fish piping common. -- Occasional fish kills. -- OTHER **4**	-- Fish avoidance or behaviors, such as erratic swimming near surface & gulping for or piping for air. More likely seen in ponds. -- Pond fish kills common. -- Frequent stream fish kills during Spring thaw. -- Very tolerant species (e.g. bullhead, catfish). -- OTHER **0**

1. Add the circled Rating Item scores to get a total for the field sheet.   TOTAL [   ]
2. Check the ranking for this site based on the total field score. Check "excellent" if the score totals at least 40. Check "good" if the score falls between 27 and 39, etc. Record your total score and rank (excellent, good, etc.) in the upper right-hand corner of the field sheet. If a Rating Item is "fair" or "poor," complete Field Sheet 4B.

RANKING	Excellent (40-46) [ ]	Good (27-39) [ ]	Fair (12-26) [ ]	Poor (11 or less) [ ]
OPTIONAL RANKING	Excellent (48-55) [ ]	Good (32-47) [ ]	Fair (14-31) [ ]	Poor (13 or less) [ ]

6. Fish behavior and condition (Refer to Field Sheet 4A, rating item 6, figure 6-2).

In addition to a degraded (less diverse) and less productive fishery, chronic sublethal doses of pesticides can lead to the following conditions, which are more likely to be observed in standing waters than in flowing waters:

- Increased susceptibility to attack by parasites and disease, such as infection by the aquatic fungus, *Saprolegnia;*

- Increased incidences of tumors;

- Behavioral changes in fish:

    - uncoordinated movements;

    - convulsive darting movements;

    - erratic swimming up and down or in a small circle;

    - sluggishness (nonresponsiveness) alternating with hyper-excitability (jumping out);

    - difficulty in respiration (adapted from Pimentel, Brown, Cross, ref. 6-4, 6-5, 6-6).

Frog tadpoles display some of the same aberrant types of behavior as fish; that is, hyper-irritability, spastic activity, and rhythmic muscular contractions that produce a whirling motion (ref. 6-9).

Chapter 7

Animal Wastes

Animal Waste Pollutants: Micro-organisms, Organic Matter, and Nutrients

Surface runoff of animal wastes contaminates a receiving body of water with four types of pollutants: (1) pathogenic and nonpathogenic micro-organisms; (2) biodegradable organic matter; (3) nutrients; and (4) salts. Ground water can be adversely affected by animal-waste nutrients and salts. Only organic matter can be seen with the naked eye, but it, too, may be degraded into fine particles that dissolve or remain suspended in the water. These particles color the water, increase its turbidity, and increase the biochemical oxygen demand (BOD). Refer to figure 7-1 for a comparison of typical BOD concentrations in municipal and agricultural wastes. Effects of the bacteria, nutrients, and salts may be observed indirectly, such as human-health effects from shellfish contamination or as eutrophication.

Micro-organisms. Animal wastes are potential sources of approximately 150 diseases. Illnesses that may be transmitted by animal manure include bacterial diseases, such as typhoid fever, gastro-intestinal disorders, cholera, tuberculosis, anthrax, and mastitis. Transmittable viral diseases are hog cholera, foot and mouth disease, polio, respiratory diseases, and eye infections (ref. 7-3).

Figure 7-1.—BOD concentrations in municipal and agricultural wastes (ref. 7-1, 7-2).

All values are BOD_5* in milligrams per liter (mg/l).

Raw domestic (municipal) sewage		200
Treated sewage with secondary treatment		20
Milking center wastes		1,500

	Influent source to a lagoon	Effluent source from a lagoon
Dairy cattle	6,000	2,100
Beef cattle	6,700	2,345
Swine	12,800	4,480
Poultry	9,800	3,430

*The determination of Biochemical Oxygen Demand as an empirical testing procedure to determine relative oxygen requirements of wastewater, effluents, and polluted waters using a 5-day incubation period.

Numerous factors influence the nature and amount of disease-producing organisms that reach waterways. Some of these factors are climate, soil types and infiltration rates, topography, animal species, animal health, and the presence of "carrier" organisms. These latter organisms carry disease-causing micro-organisms in significant numbers, but do not contract the disease themselves. Manure, applied to the land in solid or slurry form or stored in lagoons, poses varying public health hazards, depending on the distance to watercourses, nature of the soil overlying aquifers, etc. When manure is applied on hot, sunny days, harmful bacteria die quite rapidly, virtually eliminating any potential health threat. However, rain falling on freshly applied manure may yield 10,000 to 10 million bacteria per milliliter in runoff waters. The public health hazard increases if manure is applied onto frozen ground or in the rain, or if a lagoon overflows. Direct disposal into water represents a significant threat to animals, or to humans swimming in or drinking the water (ref. 7-4).

Public health departments test for the presence of *Escherichia coli* (*E. coli*) to determine if waters classified for swimming are contaminated by organic pollution. The most commonly used indicator species of organic pollution is *E. coli*. It is generally nonpathogenic and is a member of a group of fecal coliforms, bacteria that reside in the intestine of warm blooded animals, including humans. The presence of *E. coli* does not by itself confirm the presence of pathogens. Rather, it indicates contamination by sewage or animal manure and the potential for health risks. Unfortunately, there is still no easy method for distinguishing between human and animal coliform bacteria (ref. 7-5).

For this reason and because bacterial identification requires the use of sterile technique and incubation, field offices generally have not used bacteria as pollution indicators. However, those individuals interested in using bacteria as pollution indicators should refer to the last page in appendix B and to *Standard Methods for the Examination of Water and Waste Water* (ref. B-5) for details.

Organic Matter. Animal waste contaminates receiving waters with oxygen-demanding organic matter, including organic nitrogen and phosphorus compounds. When manure enters a standing water body, such as a pond, it is subject to natural decay. As decomposition occurs, BOD increases, dissolved oxygen (DO) decreases, and ammonia is released. Low DO levels and increased ammonia cause stress to stream inhabitants. Fish, in particular, are sensitive to ammonia. Nonionic (un-ionized) ammonia (NH_3) concentrations as low as 0.2 ppm may prove toxic to fish (ref. 7-6).

Animal manure is commonly spread on frozen ground in cold regions. When snowmelt runoff occurs in early spring, some of the manure washes away in the runoff from the frozen ground, contaminating nearby watercourses and bodies. Fish kills are common under these circumstances. Frequently, the receiving waters are the farm's own pond or stream.

It is only later in the spring after a complete thaw that manure nutrients are able to seep into the soil. Even then, since the crop has not been planted, or if planted, is immature and lacks extensive root systems, more than half of the nutrients can wash through the soil or run off it. Since surfaces coated with very dry or very wet manure repel water, there is greater runoff in range areas or feedlots under these conditions compared to less runoff from water-absorbing, moderately moist manured areas. In general, from 0.22 to 0.5 in of rain is necessary to produce runoff. Monitoring has shown that manure solids in late-February and early-March runoffs can be ten times more concentrated than summer rain-storm runoffs (ref. 7-3, 7-4, 7-7).

Fast-moving (lotic) waters usually can effectively degrade moderate amounts of manure and organic matter without severe declines in water quality. However, since lakes and ponds (lentic waters) are characterized by lesser flows, they often have less dissolved oxygen. They usually degrade less manure and organic matter and can be easily overloaded.

Nutrients. The effects of nutrient enrichment on receiving waters, whether nutrients come from fertilizers or manure, are the same. Since this is the case, the effects of nutrients on receiving waters discussed in Chapter 5 are applicable here.

Salts. Salts are added to animal feeds to maintain the animal's chemical balance and increase weight. Excess salts pass through the animals and are eliminated in the wastes. When manure accumulates, salt leaching becomes a potential pollution problem. With sufficient rainfall and runoff, salts can contribute to surface and ground water pollution (ref. 7-8).

Animal Waste Indicators for Receiving Waters

1. **Evidence of animal waste: visual and olfactory** (Refer to Field Sheet 2A, rating item 1, figure 7-2).

    The most obvious indication of fresh manure, even at a distance, is the unpleasant odor and the smell of ammonia. Closer visual inspection of the water and the water's edge is necessary to locate dried sludge, which may be fairly odorless.

2. **Turbidity and color** (Refer to Field Sheet 2A, rating item 2, figure 7-2).

    When manure enters water, it disintegrates fairly rapidly into small particulate matter. When the manure input is heavy and the rate of water flow is low, a noticeable increase in turbidity might occur (i.e., water may appear more opaque or cloudy).

    Nutrients contained in the manure eventually dissolve and are taken up by plants. The indirect effects of increased nutrients manifest themselves in both the vigor and amount of aquatic vegetation. For a detailed discussion of these effects, refer to chapter 5.

3. **Aquatic vegetation; fish behavior; bottom-dwelling organisms** (For rating items 3, 4, and 5 on Field Sheet 2A, see items 1, 3, and 5 in Chapter 5).

    Some of the same water-use impacts noted for nutrients in item 4, chapter 5 are also true for manure. For example, waters having excessive inputs of manure are often characterized by reduction in fishery quality. These waters also have reduced recreational use because of odors, muddy conditions, decay of massive amounts of vegetation, etc. Property near or adjacent to these waters is often devalued.

    Health effects, such as blue baby syndrome or waterborne bacterial and viral diseases sometimes occur. The closing of bacterially contaminated areas to fishing or recreation by public health agencies is sometimes due to animal waste pollution from agricultural sources. Drinking water may also be impaired by taste, color, odor, or turbidity problems.

Figure 7-2

Animal Waste

FIELD SHEET 2A: ANIMAL WASTE
INDICATORS FOR RECEIVING WATERCOURSES AND WATER BODIES

Evaluator _____ County/State _____ Date _____
Water Body Evaluated _____ Water Body Location _____ Total Score/Rank _____

Rating Item	Excellent	Good	Fair	Poor

(Circle one number among the four choices in each row which BEST describes the conditions of the watercourse or water body being evaluated. If a condition has characteristics of two categories, you can "split" a score.)

#	Rating Item	Excellent	Good	Fair	Poor
1.	Evidence of animal waste: visual and olfactory	-- No manure in or near water body. -- No odor. -- OTHER **9**	-- Occasional manure droppings where cattle cross or in water. -- Slight musk odor. -- OTHER **6**	-- Manure droppings in concentrated localized areas. -- Strong manure or ammonia odor. -- OTHER **2**	-- Dry and wet manure all over banks or in water. -- Strong manure & ammonia odor. -- OTHER **0**
2.	Turbidity & color (observe in slow water)	-- Clear or slightly greenish water in pond or along the whole reach of stream. -- No noticeable colored film on submerged objects or rocks. -- OTHER **9**	-- Occasionally turbid or cloudy. Water stirred up & muddy & brownish at animal crossings. -- Pond water greenish. -- Rocks or submerged objects covered with thin coating of green, olive, or brown build-up less than 5 mm thick. -- OTHER **6**	-- Stream & pond water bubbly, brownish and cloudy where muddied by animal use. -- Pea green color in ponds when not stirred up by animals. -- Bottom covered w/green or olive film. Rocks or submerged objects coated with heavy or filamentous build-up 5-75 mm thick or long. -- OTHER **3**	-- Stream & pond water brown to black, occasionally with a manure crust along banks. -- Sluggish & standing water—murky and bubbly (foaming). -- Ponds often bright green or with brown/black decaying algal mats. -- OTHER **0**
3.	Amount of aquatic vegetation	-- Little vegetation—uncluttered look to stream or pond. -- What you would expect for a pristine water body in area. -- Usually fairly low amounts of many different kinds of plants. -- OTHER **8**	-- Moderate amounts of vegetation; *or* -- What you would expect for the naturally occurring site-specific conditions. -- OTHER **6**	-- Cluttered weedy conditions. Vegetation sometimes luxurious and green. -- Seasonal algal blooms. -- OTHER **3**	-- Choked weedy conditions or heavy algal blooms or no vegetation at all. -- Dense masses of slimy white, greyish green, rusty brown or black water molds common on bottom. -- OTHER **0**
4.	Fish behavior in hot weather; fish kills, especially before dawn	-- No fish piping or aberrant behavior. -- No fish kills. -- OTHER **8**	-- In hot climates, occasional fish piping or gulping for air in ponds just before dawn. -- No fish kills in last two years. -- OTHER **5**	-- Fish piping common just before dawn. -- Occasional fish kills. -- OTHER **3**	-- Pronounced fish piping. -- Pond fish kills common. -- Frequent stream fish kills during spring thaw. -- Very tolerant species (e.g., bullhead, catfish). -- OTHER **0**
5.	Bottom dwelling aquatic organisms	-- Intolerant species occur: mayflies, stoneflies, caddisflies, water penny, riffle beetle and a mix of tolerants. -- High diversity. -- OTHER **9**	-- A mix of tolerants: shrimp, damselflies, dragonflies, black flies. -- Intolerants rare. -- Moderate diversity. -- OTHER **5**	-- Many tolerants (snails, shrimp, damselflies, dragonflies, black flies). -- Mainly tolerants and some very tolerants. -- Intolerants rare. -- Reduced diversity with occasional upsurges of tolerants, e.g. tube worms, and chironomids. -- OTHER **3**	-- Only tolerants or very tolerants: midges, craneflies, horseflies, rat-tailed maggots, or none at all. -- Very reduced diversity. upsurges of very tolerants common. -- OTHER **0**

1. Add the circled Rating Item scores to get a total for the field sheet.   TOTAL [   ]
2. Check the ranking for this site based on the total field score. Check "excellent" if the score totals at least 35. Check "good" if the score falls between 21 and 34, etc. Record your total score and rank (excellent, good, etc.) in the upper right-hand corner of the field sheet. If a Rating Item is "fair" or "poor," complete Field Sheet $2B_1$ or $2B_2$.

RANKING   Excellent (35-43) [   ]   Good (21-34) [   ]   Fair (7-20) [   ]   Poor (6 or less) [   ]

Chapter 8

Salts

More than 90 percent of the total irrigated land in the United States is distributed in 8 major river basins of the West, encompassing parts of 17 States (fig. 8-1). California and Texas lead the Nation in the number of irrigated acres. The major water quality problem identified in seven out of the eight basins is salinity or salt pollution (ref. 8-1). Half of the 90 to 100 million tons of salt delivered annually to watercourses comes from agriculturally related activities (ref. 8-2). Salinity is commonly measured and expressed as milligrams per liter (mg/l) or parts per million (ppm).

Approximately one-fourth of all irrigated land (about 10 million acres) suffers from salt-caused crop yield reductions (ref. 8-3). Although the most severe salt problems occur in the arid and semiarid West (fig. 8-2), increasing salinity is symptomatic of water use and reuse.

Salinity Indicators for Receiving Waters

1. Geology of area and geochemistry of water (Refer to Field Sheet 5A, rating item 1, figure 8-3).

Salts come from natural sources and result from human activities. Natural sources include geologic formations of marine origin, soils with poor drainage, salty ground water, and salty springs. The salinity of the soil is increased primarily by overapplying irrigation water to areas where drainage is inadequate. The salinity of receiving waters is increased primarily by over-irrigating lands underlain with salt-bearing layers.

Saline waters contain a number of salts whose relative proportions reflect the geology of the region as well as seasonal changes in hydrology. Consequently, salt proportions tend to be site-specific. The primary components of the dissolved solids that constitute saline water are chlorides, sulfates, and bicarbonates of the following elements: sodium, calcium, magnesium, and potassium (ref. 8-3).

Figure 8-1

Hydrologic Divisions.[1]

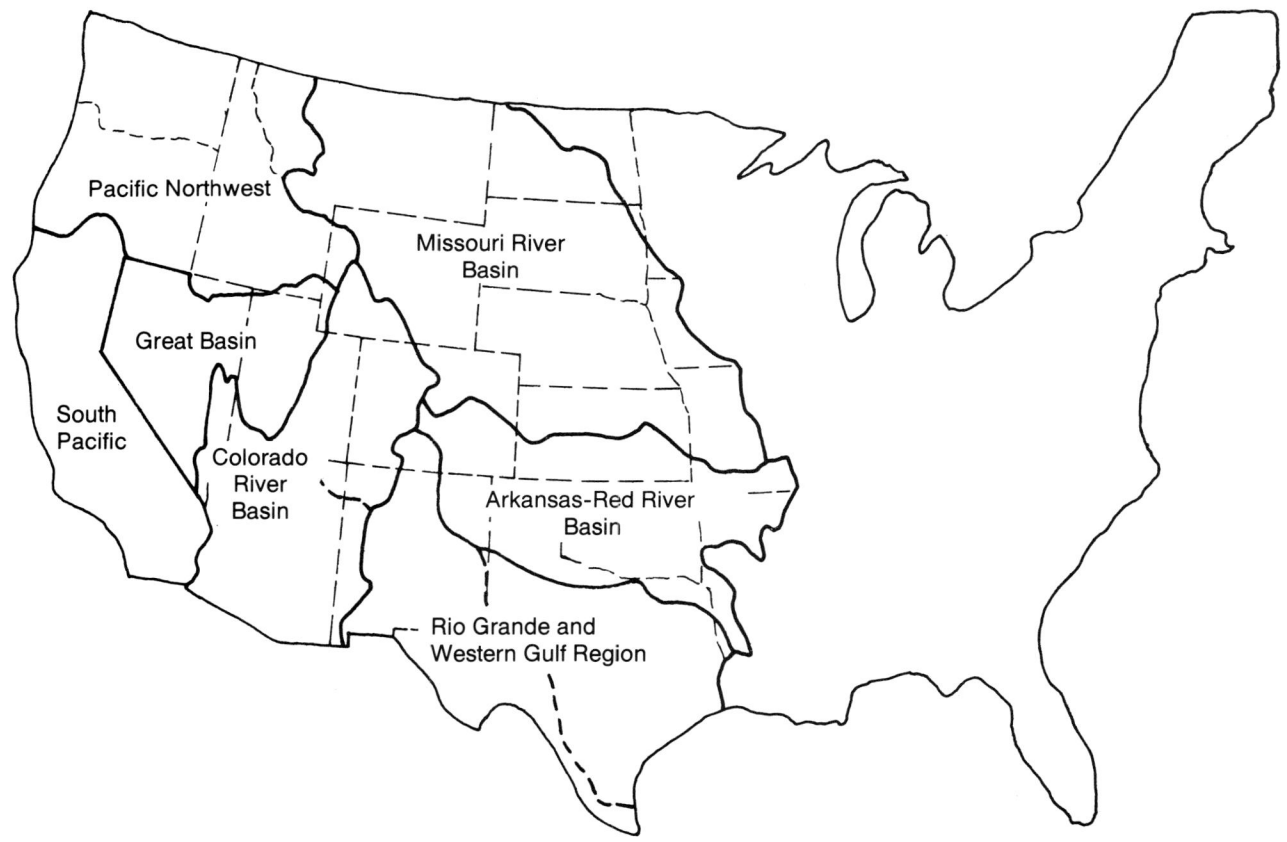

[1]Source: EPA - Pollution Control Manual for Irrigated Agriculture (Ref. 8-1).

Figure 8-2

Levels of Dissolved Solids (Mg/1) in U.S. Streams.[2]

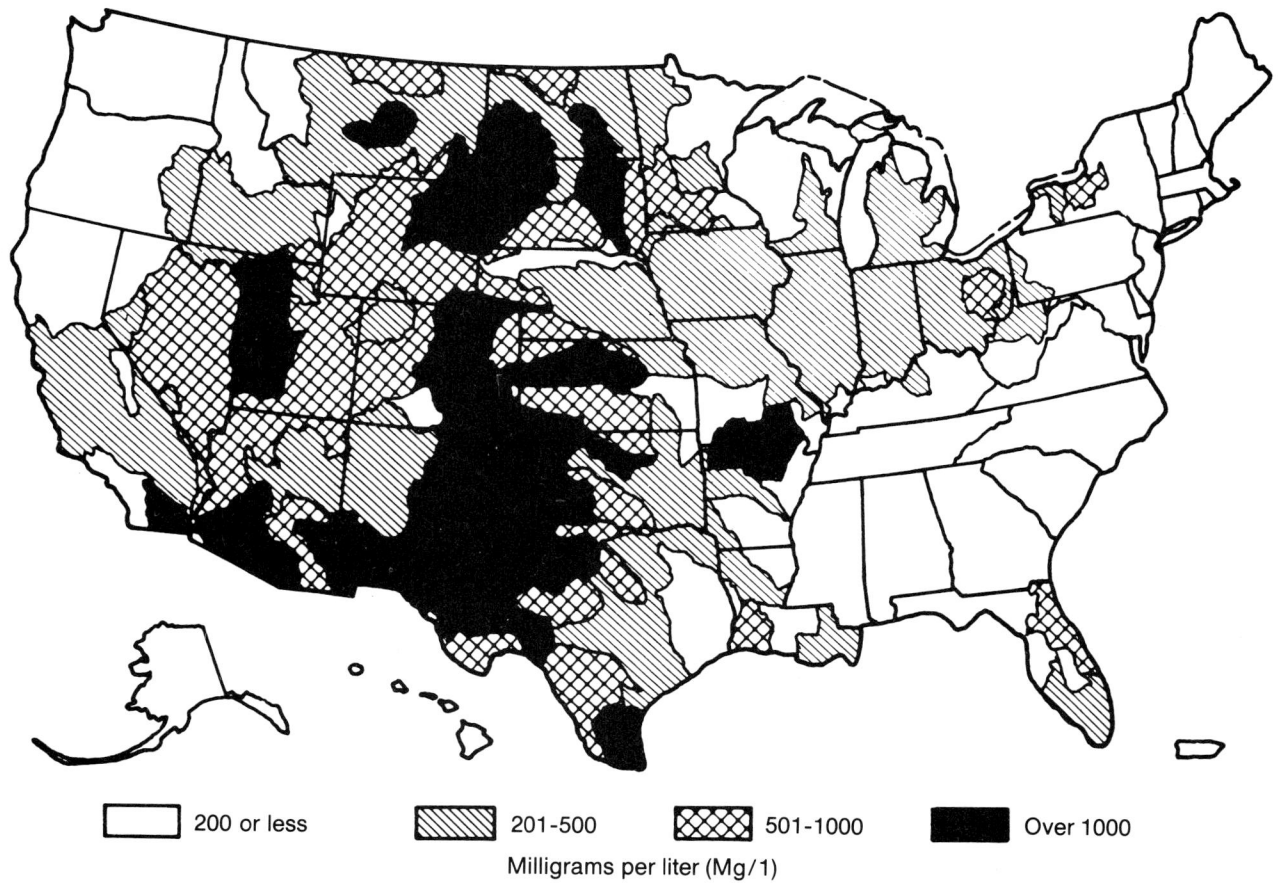

200 or less   201-500   501-1000   Over 1000

Milligrams per liter (Mg/1)

[2]Source: EPA - Council on Environmental Quality. Analysis of U.S. Geological Survey data of the National Stream Quality Accounting Network (Ref. 8-4).

2. **Precipitation and irrigation requirements** (Refer to Field Sheet 5A, rating item 2, figure 8-3).

The salinity of both water and soil is increased by salt concentration and salt loading. High salt concentrations are more likely to occur in semiarid and arid regions where evaporation exceeds precipitation. In these regions of salt-bearing layers, the usual salty water becomes even saltier as water is lost by evaporation from soil and plants (evapotranspiration). Salt pollution is even more likely to occur in these regions when drainage is inadequate or if water tables are "perched" close to the surface (5 feet or less).

Salt "loading" occurs when irrigation water percolates through a salt-laden soil profile or geologic layer on its way back to a river or stream, or when irrigation return flows accumulate salt as they run over the soil surface. The greater the irrigation requirements, the greater the opportunity for salt loading of soils.

For example, of the 10 million metric tons of salt annually reaching the Lower Colorado River Basin, 600,000 to 700,000 metric tons are annually contributed by the Grand Valley. The salt-load contribution from the Grand Valley is the result of saline subsurface irrigation return flows reaching the Colorado River. The alluvial soils of Grand Valley are high in natural salts. However, the most significant salt source is the Mancos Shale geologic formation, which underlies these alluvial soils and which contains crystalline lenses of salt that are readily dissolved by subsurface return flows (ref. 8-3, 8-5).

In this area, the irrigation water applied is at least three times greater than the crop water requirements. Although much of this excess water returns to open drains as surface runoff, having negligible effect on the Colorado River salinity, significant water quantities still reach the underlying Mancos Shale formation and pass to a near-surface cobble aquifer, where the water is returned into the Colorado River (ref. 8-5).

Figure 8-3

Salinity

FIELD SHEET 5A: SALINITY
INDICATORS FOR RECEIVING WATERCOURSES AND WATER BODIES

Evaluator _____ County/State _____ Date _____
Water Body Evaluated _____ Water Body Location _____ Total Score/Rank _____

Rating Item	Excellent	Good	Fair	Poor
	(Circle one number among the four choices in each row which BEST describes the conditions of the watercourse or water body being evaluated. If a condition has characteristics of two categories, you can "split" a score.)			
1. Geology of area and geochemistry of water	-- Agricultural area overlies formations of igneous or metamorphic origin. -- Few fractures or faults in the area. -- Very low to low mineral content—soft waters of the East and Southeast. -- OTHER **10**	-- Agricultural area primarily overlies formations of igneous or metamorphic origin with occasional areas above marine deposits. -- Few fractures or faults. -- Low to moderate mineral content—soft waters. -- OTHER **7**	-- Agricultural area overlies marine deposits. -- Faulting common. -- Moderate to high mineral content—hard waters of mountain states, deserts, and Great Plains. -- OTHER **3**	-- Agricultural area overlies marine deposits of recent origin. -- Fractures and faulting very common in the area. -- High to very high mineral content. Soils of marine origin. Salty ground water and springs. Mineral springs. Saltwater intrusion. -- OTHER **0**
2. Precipitation and irrigation requirements	-- Average crop water consumption is equal to or less than average precipitation. -- Minimal irrigation required. -- OTHER **8**	-- Average crop water consumption is between 5 & 10% more than average precipitation. -- Moderate irrigation req'd. -- OTHER **6**	-- Average crop water consumption is between 10 & 25% more than precipitation. -- Considerable irrigation required. -- OTHER **4**	-- Average crop water consumption exceeds average precipitation by more than 25%. -- Maximal irrigation required. -- OTHER **0**
3. Location of watercourse in watershed	-- Near headwaters. -- OTHER **9**	-- Not far from headwaters. -- OTHER **7**	-- Moderate distance from headwaters. -- OTHER **5**	-- Far from headwaters. -- OTHER **3**
4. Appearance of water's edge (shoreline or banks)	-- No evidence of salt crusts. -- OTHER **9**	-- Some evidence of white, crusty deposits on banks. -- OTHER **6**	-- Numerous localized patches of white, crusty deposits on banks. -- OTHER **4**	-- Most of the pond or stream bank covered with a thick, white, crusty deposit. Salt "feathering" on posts abundant. -- OTHER **1**
5. Appearance of aquatic vegetation	-- No evidence of wilting, toxicity, or stunting. -- OTHER **10**	-- Minimal wilting and toxicity, bleaching, leaf burn. -- Little if any stunting. -- OTHER **7**	-- Stream or pond vegetation may show wilted and toxic symptoms—bleaching, leaf burn. -- Presence of some salt-tolerant species. -- OTHER **3**	-- Evidence of severe wilting, toxicity, or stunting. -- Presence of only the most salt-tolerant species or complete absence of vegetation. -- OTHER **0**
6. Streamside vegetation	-- Very few species. -- OTHER **8**	-- Few salt tolerant species. Refer to list below*. -- OTHER **7**	-- Increasing dominance of salt-tolerant species. -- OTHER **5**	-- Vegetation almost totally salt-tolerant species or absence of vegetation. -- OTHER **3**
OPTIONAL 7. Animal teratology (birth defects & tumors in fish and other animals)	-- No birth defects or tumors. -- OTHER **10**	-- Minimal birth defects & tumors occuring in the population randomly. -- OTHER **6**	-- Some birth defects & tumors. -- OTHER **1**	-- Considerable numbers of birth defects & tumors. -- OTHER **0**

*Salt-tolerant species include greasewood, alkali sacaton, fourwing saltbush, shadscales, saltgrass, tamarisk (salt cedar), galleta, western wheatgrass, crested wheat, mat saltbush, reed canarygrass, and rabbitbrush.

1. Add the circled Rating Item scores to get a total for the field sheet. TOTAL [   ]
2. Check the ranking for this site based on the total field score. Check "excellent" if the score totals at least 47. Check "good" if the score falls between 32 and 46, etc. Record your total score and rank (excellent, good, etc.) in the upper right-hand corner of the field sheet. If a Rating Item is "fair" or "poor," complete Field Sheet $5B_1$ or $5B_2$.

RANKING	Excellent (47-54) [ ]	Good (32-46) [ ]	Fair (15-31) [ ]	Poor (14 or less) [ ]
RANKING (optional)	Excellent (55-64) [ ]	Good (35-54) [ ]	Fair (16-34) [ ]	Poor (15 or less) [ ]

Figure 8-4

Salinity

FIELD SHEET 5B$_2$. SALINITY
INDICATORS FOR SALINE SEEPS

Evaluator _____ County/State _____ Date _____ : Practices
Saline Seep Evaluated _____ Seep Location _____ Total Score/Rank _____ : from
Rating Item : Excellent : Good : Fair : Poor : Appendix E

(Circle one number among the four choices in each row which BEST describes the conditions of the field or area being evaluated. If a condition has characteristics of two categories, you can "split" a score.)

Rating Item	Excellent	Good	Fair	Poor	Practices from Appendix E
1. Geology	-- Agricultural area overlies formations of igneous or metamorphic origin. -- Few fractures or faults in the area. -- OTHER **10**	-- Agricultural areas primarily overlies formations of igneous or metamorphic origin with occasional areas above marine deposits. -- Few fractures or faults. -- OTHER **7**	-- Agricultural area overlies marine deposits. -- Faulting common. -- OTHER **3**	-- Agricultural area overlies marine deposits of recent origin. -- Fractures and faulting very common in the area. -- OTHER **0**	
2. Precipitation and irrigation requirements	-- Average crop water consumption is equal to or less than average precipitation. -- OTHER **8**	-- Average crop water consumption is between 5 and 10% more than average precipitation. -- OTHER **6**	-- Average crop water consumption is between 10 and 25% more than precipitation. -- OTHER **4**	-- Average crop water consumption exceeds average precipitation by more than 25%. -- OTHER **0**	
3. Cropping system	-- Crop rotation consists of annual crops with maximum consumptive water use. -- OTHER **8**	-- Crop rotation consists of annual crops. -- OTHER **6**	-- Crop rotation contains a biannual fallow period. -- Crops with maximum water consumptive use grown. -- OTHER **4**	-- Crop rotation contains a biannual fallow period. -- OTHER **2**	17,37,68, 72
4. Field appearance, including salt crusts	-- Downslope fields exhibit even-appearing crop growth. High yields are common for the area. -- OTHER **9**	-- Downslope areas of field or downslope fields exhibit even crop growth, but of reduced yield for the area. -- OTHER **7**	-- Downslope areas of field or downslope fields have uneven growth of crops with patches of crops significantly stunted. -- Occasionally white crust occurs in these patches. -- OTHER **3**	-- Downslope areas of fields have bare spots not accounted for by soil variations. Bare spots occur in highly saline soils. White crust common. -- OTHER **1**	

1. Add the circled Rating Item scores to get a total for the field sheet. TOTAL [   ]
2. Check the ranking for this site based on the total field score. Check "excellent" if the score totals at least 30. Check "good" if the score falls between 20 and 29, etc. Record your total score and rank (excellent, good, etc.) in the upper right-hand corner of the field sheet. If a Rating Item is "fair" or "poor," find the practices in the right-hand column to help remedy the conditions.

RANKING   Excellent (30-35) [   ]   Good (20-29) [   ]   Fair (8-19) [   ]   Poor (7 or less) [   ]

3. **Location of watercourses in watershed** (Refer to Field Sheet 5A, rating item 3, figure 8-3).

   In geologic regions where the soils are underlain by salt-bearing layers, the salinity of receiving watercourses increases with the distance from the headwaters. The salinity is least near the headwaters, where there has been little opportunity for salt loading or salt concentration, and greatest downstream, where effects of these two processes are maximized. Generally, salt loading is the major cause of salinity increases in the arid and semiarid regions of the United States. Salinity in the Colorado River ranges from an average of less than 50 milligrams per liter (mg/l) in the headwaters to 825 mg/l at Imperial Dam and 950 mg/l in Mexico (ref. 8-3, 8-6, 8-7).

4. **Appearance of water's edge (shoreline or banks)** (Refer to Field Sheet 5A, rating item 4, figure 8-3).

   The most obvious indicator of excessive salinity is the presence of white, crusty deposits of salts. These deposits may occur at the high water mark along the banks of a stream or river, or at seepage points along a high bank or cliff. "Salt feathering," the crystallizing of salt in feathery-like patches on posts and tree stumps, is another indicator of highly saline conditions.

5. **Appearance of aquatic vegetation** (Refer to Field Sheet 5A, rating item 5, figure 8-3).

   Salt pollution becomes a problem when the concentration of salts in the soil/water solution interferes with the growth of plants. Table salt (sodium chloride) is often the dominant salt present. It affects plants in two ways: (1) By increasing

Figure 8-5

Generalized Diagram of Saline Seep, Recharge Area, and the Substrata Formation That Contributes to a Saline Seep (Ref. 8-10).

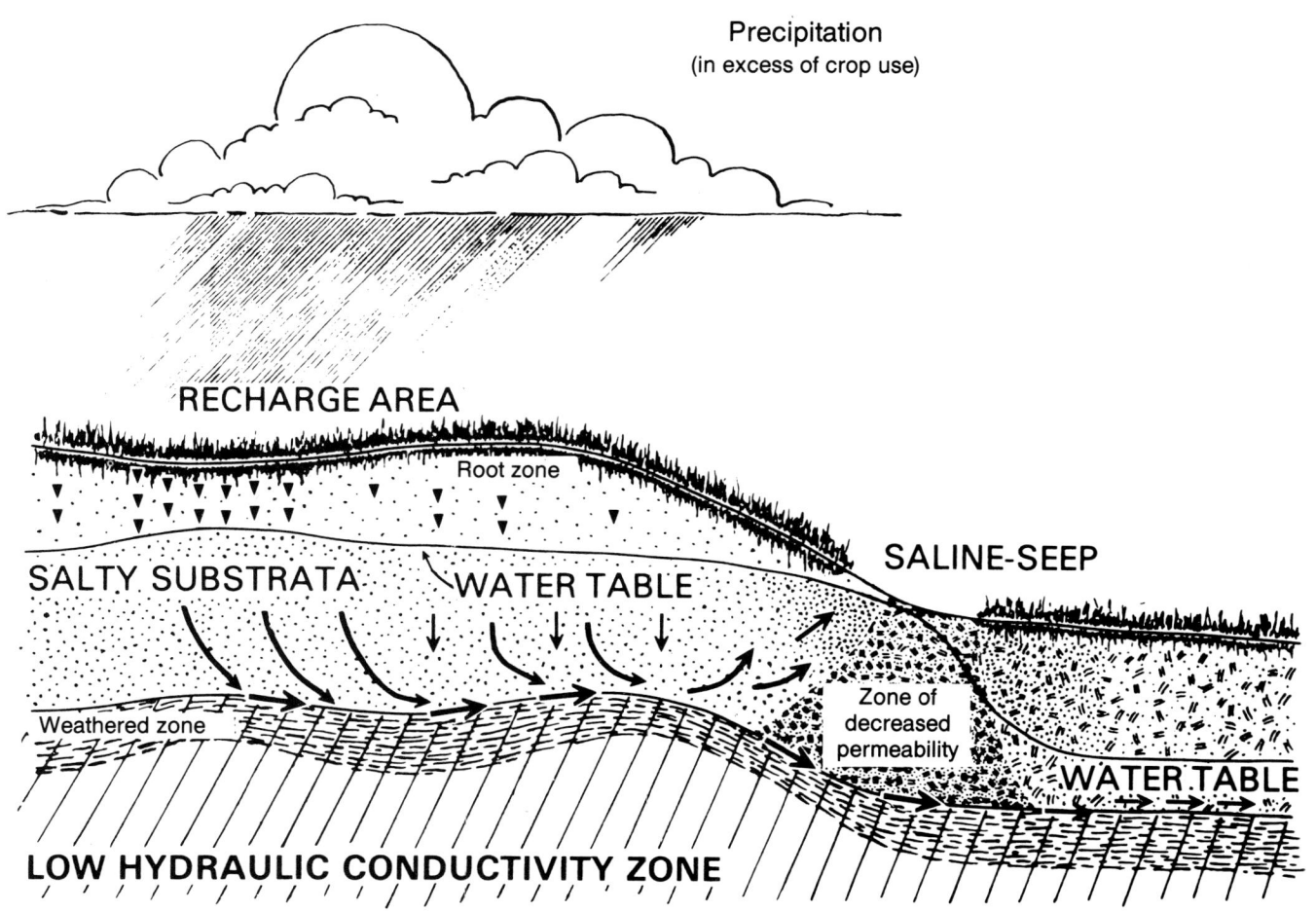

the osmotic pressure, it reduces the amount of water that plants can take up, leading to stunted growth and reduced yields; (2) At high concentrations it causes toxic effects, such as leaf tip and marginal leaf burn, chlorosis (bleaching), or defoliation (ref. 8-8).

6. **Streamside vegetation** (Refer to Field Sheet 5A, rating item 6, figure 8-3).

    As the salinity of water increases, salt-intolerant species die and are replaced by more salt-tolerant types. Examples of the latter are greasewood, alkali sacaton, fourwing saltbush, shadscales, saltgrass, tamarisk (salt cedar), galleta, western wheatgrass, mat saltbush, reed canarygrass, and rabbitbrush. Some emergent rooted aquatics, such as cattails, appear to be tolerant of even the highest concentration of salts.

7. **Animal teratology (birth defects)** (Refer to Field Sheet 5A, rating item 7, figure 8-3).

    Severe toxic effects and birth defects or tumors in animals have been observed in isolated areas (e.g., Kesterson Reservoir in California) because of high concentrations of toxic compounds, selenium, or flouride. Some newly hatched ducks and other birds in the Kesterson Reservoir, which had elevated levels of selenium, lacked ears, eyes, beaks, wings, or legs (ref. 8-9).

8. **Salinity indicators** (Refer to Field Sheet $5B_2$, figure 8-4. Field Sheet $5B_2$ should only be used in areas where the geology makes saline seeps possible.)

    A white salty crust can be an indicator of a saline seep. Saline seeps are in those localities where saline water surfaces downslope of its recharge area. The seeping water results from excess root zone moisture that percolates through salt-bearing layers. Water, leaching below the root zone, carries dissolved salt to the surface downslope of the area of infiltration. These areas are common in the Northern Great Plains Region (Montana, North Dakota, and South Dakota), where precipitation percolates through salt-laden glacial till into ground water, emerging later in a discharge area at another location (fig. 8-5).

    Indicators for saline seeps are land-based, not water-based. Rating items 1 and 2 approximate those discussed above in field sheet 5A, "Salinity Indicators for Receiving Waters." Other indicators include the type of cropping system (rating item 3) and the appearance of field crops downslope of the recharge area (rating item 4). Saline seep areas will have uneven growth of crops with some significantly stunted patches or bare spots. White salt crusts occur in occasional patches in areas considered to be "fair," and are common under "poor" conditions.

    Since saline seeps result from excess moisture in the soil profile, it is important to consider the cropping system thoroughly. There will be less "seeping" when crops with the maximum consumptive water use are planted. This is especially true when crops are grown on an annual basis; i.e., when the fields are not allowed to lie fallow.

APPENDIX A

Water Quality Procedures

- Sequential Comparison Index
- Beck's Biotic Index
- Floating Body Technique

Sequential Comparison Index

The Sequential Comparison Index (SCI) is a simple stream quality method, based upon distinguishing organisms by color, size, and shape, and requires no taxonomic expertise (ref. 8-4). The only needs are to be able to distinguish the number of different types (taxa) of organisms and the number of "runs" in samples containing less than 250 organisms. A diversity index (DI) is obtained by dividing the number of runs by the number of specimens. This index is multiplied by the number of taxa to give the final DI. DI values of 12 or above are indicative of healthy streams with high diversity and a balanced density. Polluted streams typically have DI values of 8 or less.

Sample analysis. There are many methods of biological specimen analysis. Diversity indices are useful because they condense considerable data into a single numerical value. The SCI is a simple diversity method which can be used by a non-biologist. The following is a brief summary of the SCI evaluation. For detailed information see Cairns' article on simple biological assessment (ref. 5-5).

Bottom sample collection and preservation. Bottom samples from a watercourse or water body should be collected with an appropriate sampler. If a bottom sampler is not available, trowels or shovels can be used to collect the sample. Place the material collected into a tub or bucket. Dilute the material with water and swirl. Pour it through a U.S. Standard #30 sieve or a 30-mesh screen. Remove rocks, sticks, and other artifacts after carefully checking for clinging organisms. Wash the screened material into a container and preserve it in 10 percent formalin or 70 percent ethanol (ethyl alcohol). Organisms may be sorted from the sample detritus in the field with forceps or at the laboratory. It is often desirable, prior to preserving the sample, to place rocks, sticks, and other objects in a white pan partially filled with water. Many of the animals will float free from the objects or can be removed with forceps. All samples should be stored in a suitable container and preserved with 10 percent formalin or 70 percent ethanol. The samples should be labeled with the location, date, type of sampler used, name of collector, and other pertinent information.

SCI Procedure.

1. Randomize specimens in a jar by swirling.
2. Pour specimens into a lined white enamel pan.
3. Disperse clumps of specimens by pouring preservative or water on the clumps.
4. Determine the number of runs in the sample by comparing two specimens at a time. The current specimen need only be compared with the previous one. If it is similar, it is part of the same run. If not, it is part of a new run (fig. A-1). A 2X magnifying glass or a low-powered binocular microscope is needed for this operation.
5. Determine the total number of specimens in the sample.
6. Calculate DI_1:

$$DI_1 = \frac{\text{number of runs}}{\text{number of specimens}}$$

7. Record the number of different taxa observed. This does not require a specialist in taxonomy. Most bottom fauna organisms are fairly easily divided into recognizable entities by non-biologists. *Identification of the organisms is not necessary.* A 2X magnifying glass or a low powered binocular microscope is needed for this operation.

8. Determine from figure A-2 the number of times (N) the SCI examination must be repeated on the sample to be 95 percent confident that the mean DI_1 is within a desired percentage of the true value for DI_1. In most pollution work involving gross differences, line A of figure A-2 should be used.

9. After determining N from figure A-2, rerandomize the sample and repeat the SCI examination on the same sample N-1 times. Calculate the average DI_1 by the following equation:

$$\overline{DI_1} = \frac{\Sigma DI_1}{N}$$

10. DI_T is a diversity index value. It represents species diversity and, therefore, health of a watercourse. Calculate DI_T by the following equation:

$$DI_T = \overline{DI_1} \times \text{No. of Taxa}$$

11. Repeat the above procedure for each sample collection.

The above procedure should only be used on samples containing fewer than 250 specimens. Healthy streams with a high diversity and a balanced density tend to have DI_T values above 12. Polluted communities tend to have DI_T values of 8 or less. Intermediate values have been found in semipolluted streams.

To determine if different bottom-fauna community structures are significantly different from each other, calculate the 95 percent confidence intervals around each DI_T value. If the intervals do not overlap, then the community structures are significantly different. For example, if sampling station "A" has a DI_T value of 25, station "B" has a DI_T value of 10, and line A of figure A-2 is used, then the 95 percent confidence interval would be 20 percent, or 10 percent on either side of the determined DI_T value. Station "A" has 95 percent confidence interval for the DI_T value from 22.5 to 27.5 (20 percent of 25 = 5). Station "B" has a 95 percent confidence interval for the DI_T value from 9 to 11 (20 percent of 10 = 2). The 95 percent confidence intervals do not overlap, and therefore the bottom fauna communities at the two stations are significantly different.

The SCI examination is a useful tool. It requires no taxonomic expertise. It is easy to perform and produces results quickly. It should not be used to represent or replace other more accurate techniques requiring a person trained in aquatic biology.

Figure A-1

Determination of Runs in Sequential Comparison Index

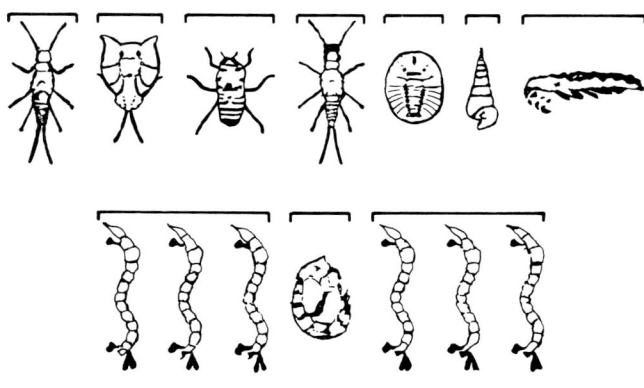

Figure A-2

Confidence Limits for DI_1 Values.

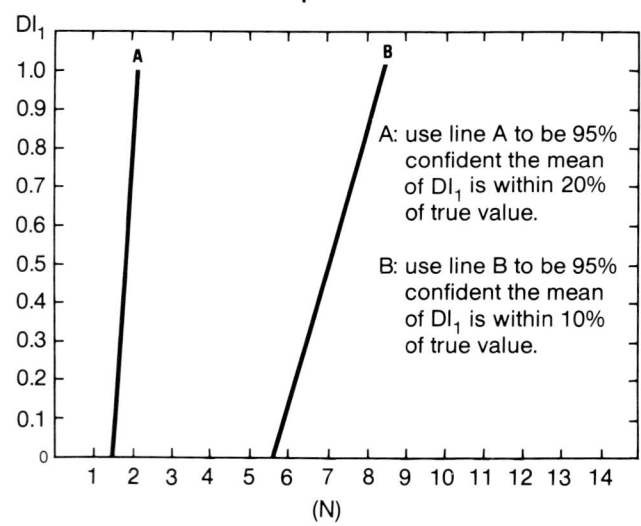

A: use line A to be 95% confident the mean of DI_1 is within 20% of true value.

B: use line B to be 95% confident the mean of DI_1 is within 10% of true value.

Beck's Biotic Index

Beck's Biotic Index (ref. 3–10) was developed primarily for use in Florida and assumes taxonomic expertise, but it can be used with generic level identification when less sensitivity is acceptable. This system can be used to indicate both the magnitude and probable cause of environmental stress. Beck developed the methodology to categorize stream macro-invertebrates (large animals without backbones, ref. A-1).

Three categories (table A-3, fig. A-4) are defined below:

Class I Organisms (Sensitive or Intolerant)
Organisms that exhibit a rapid response to aquatic environmental changes and are killed, driven out of the area, or as a group are substantially reduced in number when their environment is degraded.

Class II Organisms (Facultative)
Organisms that have the capability to live under varying conditions; e.g., a facultative anaerobe is an organism that although usually and normally lives in the presence of free oxygen, can live in absence of free oxygen. Most survive in areas where organic pollution is producing eutrophication or "enrichment" of the aquatic ecosystem.

Class III Organisms (Tolerant)
Organisms capable of withstanding adverse conditions within the aquatic environment.

According to this approach, which assumes that there are not naturally occurring limiting factors, an undisturbed community will include representatives of the majority of the groups contained in Class I as well as some representatives of Classes II and III. By contrast, a sample which consists mainly of Class II organisms is being "limited" or impacted by either natural factors, such as low flow, homogenous substrate, etc. or is impacted due to human activities. Waters dominated by Class III organisms are probably adversely affected by organic pollution.

The structure of the benthic (bottom) invertebrate community in waterways polluted by organic waste differs quantitatively from invertebrate communities in unpolluted waterways. That is, organic pollution results not just in a reduction in species richness (the total number of benthic groups), but also in a stimulation in density (the total number of organisms collected per sample).

By contrast, waterways impacted with toxic materials, such as pesticides or acid mine drainage, show decreases both in richness and density. Sediment causes a greater reduction in density than in richness. Because of the above differences and because there are often dominant organisms characteristic of sediment pollution, it is possible to differentiate sediment stress from the stresses of toxic materials and organic wastes.

For this type of investigation, a dip net is used to take a "kick" sample, which is sufficient to obtain a representative sample of the organisms present. With this procedure the net is placed upright on the bottom in an area of swift water, and the stream bottom upstream of the net is sufficiently disturbed to dislodge any organisms located there. The dislodged organisms will be carried by the current into the net and captured. Any rocks that can be overturned should be turned and any clinging organisms collected.

Mathematical expression.

$$BI = 2n_I + n_{II}$$

where:

BI = Beck's Biotic Index
n_I = the number of Class I species identified from the samples
n_{II} = the number of Class II species identified from the samples

Table A-3.—Benthic macroinvertebrates classed according to Beck's Biotic Index Classes (ref. A-2)

Invertebrate Form	Class
Caddisflies: Trichoptera	
Hydropsychidae	1
Hydroptilidae	1
Limnephilidae	1
Leptoceridae	1
Helicopsychiade	1
Psychomyiidae	1
Goeridae	1
Stoneflies: Plecoptera	
Perlidae	1
Perlodidae	1
Mayflies: Ephemeroptera	
Baetidae	1
Heptageniidae	1
Ephemeridae	1
Helligrammites	
Corydalidae	1
Freshwater Naiads (Clams)	
Unionidae	1
Beetles: Coleoptera	
Elmidae (Riffle Beetle)	1
Psephenidae (Water Penny)	1
Damselflies: Odonata	
Coenagrionidae	2
Agrionidae	2
Dragonflies: Odonata	
Aeschnidae	2
Comphidae	2
Libellulidae	2
Crayfish	
Astacidae	2
Flatworms	
Planaridae	2
Crane Flies	
Tipulidae	2
Gill Snails	
Pleuroceridae	2
Horse Flies	
Tabanidae	2
Isopods	
Asellidae (Aquatic Sowbugs)	2
Blackflies	
Simuliidae	2
Air-Breathing Snails	
Physidae	3
Ancylidae (Limpets)	3
Aquatic Earthworms	
Oligochaeta	3
Midges	
Chironomidae	3
Leeches	
Hirundinea	3
Moth flies	
Psychodidae	3

Recommended Level of Taxonomic Identification. This index should be used in conjunction with species level identification to enhance the sensitivity of the index in detecting ecosystem perturbations. The use of generic level identification requires the assignment of a tolerance classification to a genus, corresponding to the most tolerant species within that genus, and leads to decreased index sensitivity. Generic level identification can be utilized when less sensitivity is acceptable or when species identification is not possible. For example, species taxonomy within the *Chironomidae* (midges) can be so difficult as to preclude its use.

Geographic Applicability. This index has not been widely employed outside of the State of Florida.

Computational Devices Required. A simple desk-top calculator is recommended for the calculation of values.

Statistical Evaluation. Statistical evaluation of index values can be inappropriate or present interpretation difficulties. Tests of raw data (Chi square, correlation, t-test, etc.) are recommended.

Use of the index.

1. *Level of sampling required.* Sample size can be limited, depending on the degree of organic pollution encountered. Organically polluted conditions demand more extensive and precise collection and analysis of data to ensure that sensitive animals have not been overlooked.

2. *Recommended form of data reduction.* Absolute estimates of generic and/or specific representation should be entered directly into the computational formula.

3. *Modes of data display.* Index values can be displayed in either tabular and/or graphical form in a site or locale-specific manner.

4. *Interpretation.* Index values will range from 0 to approximately 40; the lower the index value, the greater the organic stress. See table below. An index value of 10 is the lowest value accepted as indicative of clean water without additional discussion.

Figure A-4

Macroinvertebrates According to Beck's Biotic Index Classes (Ref. A-2).

1. Intolerant (sensitive) to pollution:

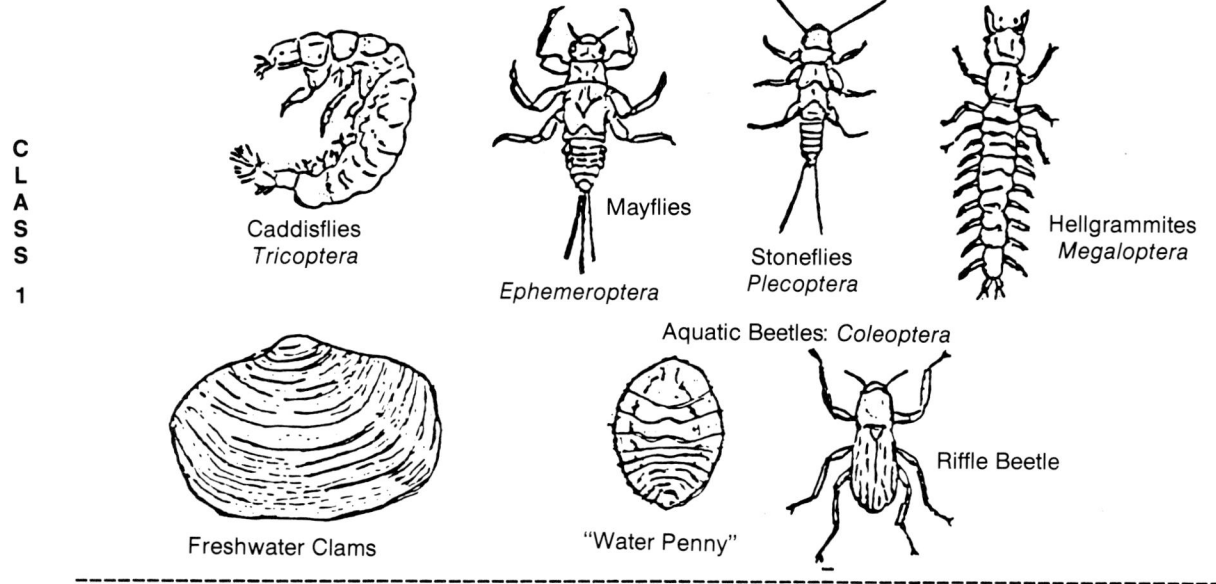

2. Facultative - Can tolerate some pollution:

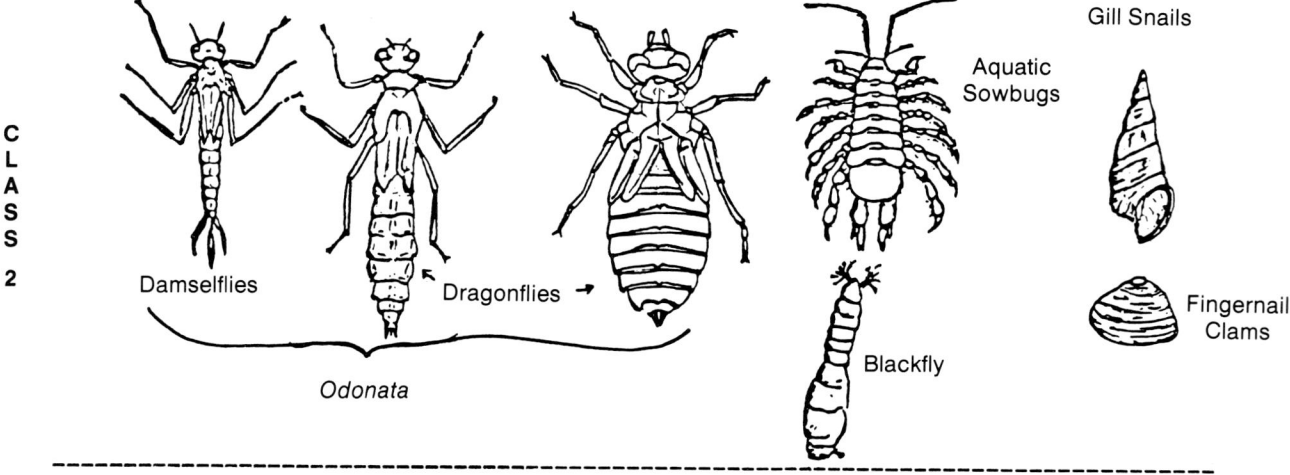

3. Tolerant to pollution:

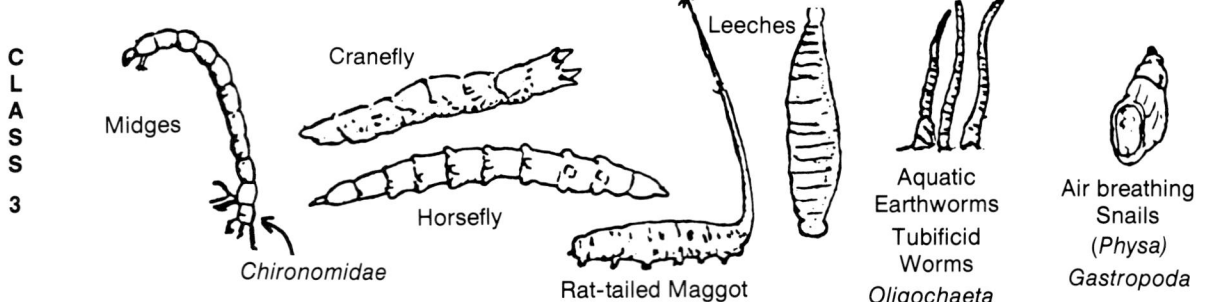

Beck's Biotic Index Values

Index Value	Description
0	Stream grossly polluted
1 to 5	Stream moderately polluted
6 to 9	Stream clean, but with a monotonous habitat and instream velocity
10 or greater	Stream clean

Floating Body Technique

The Floating Body Technique measures water flow velocity, which is calculated by measuring the time taken for a marker to travel a known distance downstream.

Procedure. A stretch of a watercourse should be selected which is approximately straight. The compass direction of this stretch should be measured. Using the compass direction, a 90 degree angle is laid out so it crosses the stream. This is most conveniently done by locating a landmark on the distant side of the stream, and moving up- or downstream to locate (and mark) the point at which a 90 degree angle exists. A distance should then be measured downstream to the other end of the straight stretch, and a similar 90 degree angle laid out. The marker or dye is tossed into the stream above the initial point and then timed to see how long it takes to get from one point to the other. If dye is used, the time is measured until the front part of the dye stain arrives. Velocity is calculated as distance divided by time.

When the velocity measured is the peak velocity of the stream (usually at the surface in the center), it is possible to calculate an approximate average for velocity for the stream, assuming typical cross section. A common average is 85 percent of the maximum current velocity.

Accuracy. This technique can be moderately accurate. The major sources of error are caused by the marker floating out of the desired path. If the maximum current velocity is desired, the marker may tend to end up in eddies along the way, rather than staying in the maximum velocity portion of the stream. This source of error can be reduced by making repeated measurements or by using dye as the marker. Calculations of average stream velocity from a measured maximum velocity are in error if the correction factor is inappropriate. Deviations from the 85 percent factor mentioned above are common.

Application Notes. This technique is inexpensive. A crew size of one is suitable for slow moving streams, but a crew size of two is necessary to signal when the marker has passed the ending point if the stream moves too fast for one crew member to move from the starting point to the ending point. It is most appropriate where streams are relatively large and have a smooth slope.

APPENDIX B

Aquatic Organisms

Algae Important in Water Supplies (ref. B–1).
- Taste and odor algae
- Filter clogging algae
- Polluted water algae
- Clean water algae
- Plankton and other surface water algae
- Algae growing on reservoir walls

Types of Freshwater Algae (ref. B–2).

Simple Assessment of Bottom-Dwelling Insects (ref. A–2).

SCS Key to the Major Invertebrate Species of Stream Zones (ref. B–3).

Diagrams of Common Fish Species (ref. B–4).

Detection of *Escherichia coli* in water samples (ref. B–5).

Table B–1.—Taste and odor algae.

Species Names	Linear Magnifications
Anabaena plactonica	250
Anacystis cyanea	250
Aphanizomenon flos-aquae	250
Asterionella gracillima	250
Ceratium hirundinella	250
Dinobryon divergens	250
Gomphosphaeria lacustris, kuetzingianum type	500
Hydrodictyon reticulatum	10
Mallomonas caudata	500
Nitella gracilis	1
Pandorina morum	500
Peridinium cinctum	500
Staurastrum paradoxum	500
Synedra ulna	250
Synura uvella	500
Tabellaria fenestrata	250
Uroglenopsis americana	125
Volvox aureus	125

Table B–2.—Filter clogging algae

Species Names	Linear Magnifications
Anabaena flos-aquae	500
Anacystis dimidiata	1000
Asterionella formosa	1000
Chlorella pyrenoidosa	5000
Closterium moniliferum	250
Cyclotella meneghiniana	1500
Cymbella ventricosa	1500
Diatoma vulgare	1500
Dinobryon sertularia	1500
Fragilaria crotonensis	1000
Melosira granulata	1000
Navicula graciloides	1500
Oscillatoria princeps (top)	250
Oscillatoria chalybea (middle)	250
Oscillatoria splendida (bottom)	500
Palmella mucosa	1000
Rivularia dura	250
Spirogyra porticalis	125
Synedra acus	500
Tabellaria flocculosa	1500
Trachelomonas crebea	1500
Tribonema bombycinum	500

Table B–3.—Polluted water algae.

Species Names	Linear Magnifications
Agmenellum quadriduplicatum, tenuissima type	1000
Anabaena constricta	500
Anacystis montana	1000
Arthrospira jenneri	1000
Carteria multifilis	2000
Chlamydomonas reinhardi	1500
Chlorella vulgaris	2000
Chlorococcum humicola	1000
Chlorogonium euchlorum	1500
Euglena viridis	1000
Gomphonema parvulum	3000
Lepocinclis texta	500
Lyngbya digueti	1000
Nitzschia palea	2000
Oscillatoria chlorina (top)	1000
Oscillatoria putrida (middle)	1000
Oscillatoria lauterbornii (bottom)	1000
Phacus pyrum	1500
Phormidium autumnale	500
Pyrobotrys stellata	1500
Spirogyra communis	250
Stigeoclonium tenue	500
Tetraedron muticum	1500

Figure B-1

Algae Important in Water Supplies.

Taste and Odor Algae

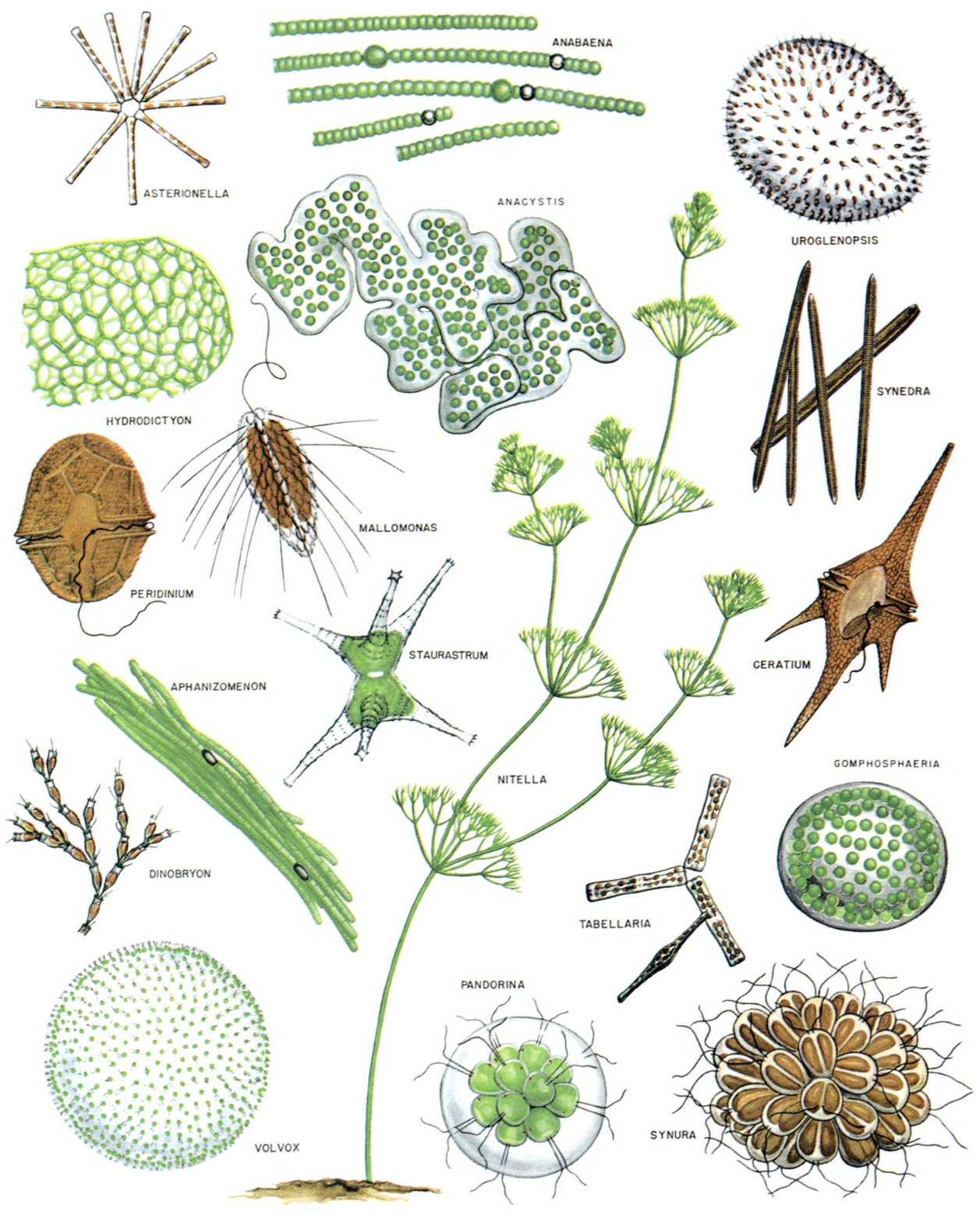

Figure B-2

Filter Clogging Algae.

Figure B-3

Polluted Water Algae.

Figure B-4

Clean Water Algae.

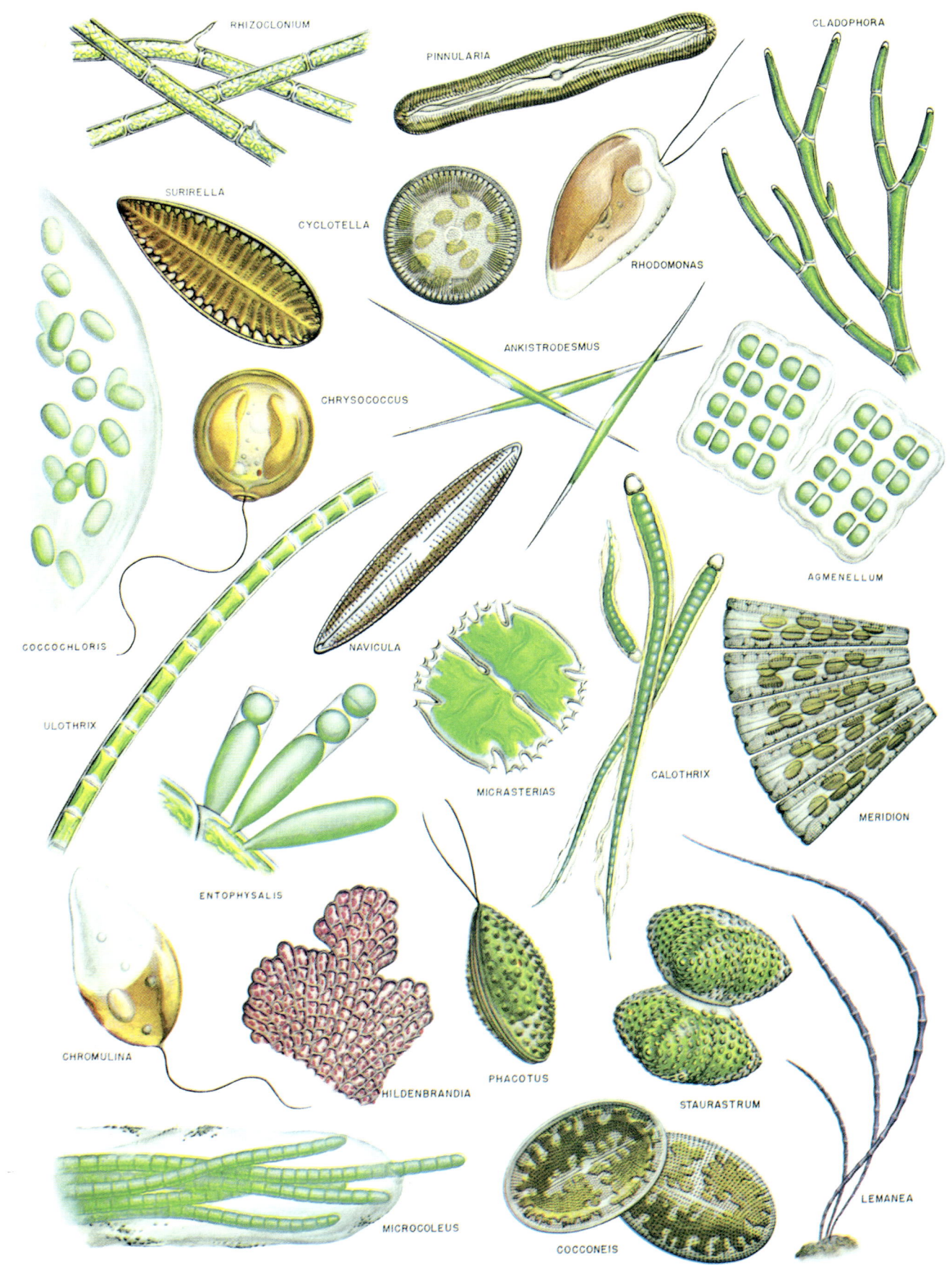

Table B-4.—Clean water algae

Species Names	Linear Magnifications
Agmenellum quadriduplicatum, glauca type	250
Ankistrodesmus falcatus var. acicularis	1000
Calothrix parietina	500
Chromulina rosanoffi	4000
Chrysococcus rufescens	4000
Cladophora glomerata	100
Coccochloris stagnina	1000
Cocconeis placentula	1000
Cyclotella bodanica	500
Entophysalis lemaniae	1500
Hildenbrandia rivularis	500
Lemanea annulata	1
Meridion circulare	1000
Micrasterias truncata	250
Microcoleus subtorulosus	500
Navicula gracilis	1000
Phacotus lenticularis	2000
Rinnularia nobilis	250
Rhizonclonium hieroglyphicum	250
Rhodomonas lacustris	3000
Staurastrum punctulatum	1000
Surirella splendida	500
Ulothrix aequalis	250

Table B-5.—Plankton and other surface water algae.

Species Names	Linear Magnifications
Actinastrum gracillimum	1000
Botryococcus braunii	1000
Coelastrum microporum	500
Cylindrospermum stagnale	250
Desmidium grevillei	250
Euastrum oblongum	500
Eudorina elegans	250
Euglena gracilis	1000
Fragilaria capucina	1000
Gomphosphaeria aponina	1500
Gonium pectorale	500
Micractinium pusillum	1000
Mougeotia scalaris	250
Nodularia spumigena	500
Oocystis borgei	1000
Pediastrum boryanum	125
Phacus pleuronectes	500
Scenedesmus quadricauda	1000
Sphaerocystis schroeteri	500
Stauroneis phoenicenteron	500
Stephanodiscus hantzschii	1000
Zygnema sterile	250

Table B-6.—Algae growing on reservoir walls.

Species Names	Linear Magnifications
Achnanthes microcephala	1500
Audouinella violacea	250
Batrachospermum moniliforme	3
Bulbochaete insignis	125
Chaetophora elegans	250
Chara globularis	4
Cladophora crispata	125
Compsopogon coeruleus	125
Cymbella prostrata	250
Draparnaldia glomerata	125
Gomphonema geminatum	250
Lyngbya lagerheimii	1000
Microspora amoena	250
Oedogonium suecicum	500
Phormidium uncinatum	250
Phytoconis botryoides	1000
Stigeoclonium lubricum	250
Tetraspora gelatinosa	125
Tolypothrix tenuis	500
Ulothrix zonata	250
Vaucheria sessilis	125

Table B-6a. — Estuarine pollution algae

Species Names	Linear Magnifications
Agardhiella tenera	2
Amphidinium fusiforme	1500
Asterionella japonica	500
Chaetoceros decipiens	750
Chaetomorpha aerea	125
Codium fragile	0.25
Enteromorpha intestinalis	0.25
Eutreptia viridis	750
Melosira sulcata	500
Nannochloris atomus	2500
Nitzschia closterium	500
Pelvetia fastigiata	0.25
Peridinium trochoideum	1000
Porphyra atropurpurea	0.25
Prasiola stipitata	10
Prorocentrum micans	750
Rhodoglossum affine	0.5
Scytosiphon lomenteria	0.25
Skeletonema costatum	500
Spirulina major	2500
Stephanoptera gracilis	2000
Stichococcus marinus	125
Trichodesmium erythraeum	2000
Ulva Iactuca	0.5

Figure B-5

Plankton and Other Surface Water Algae.

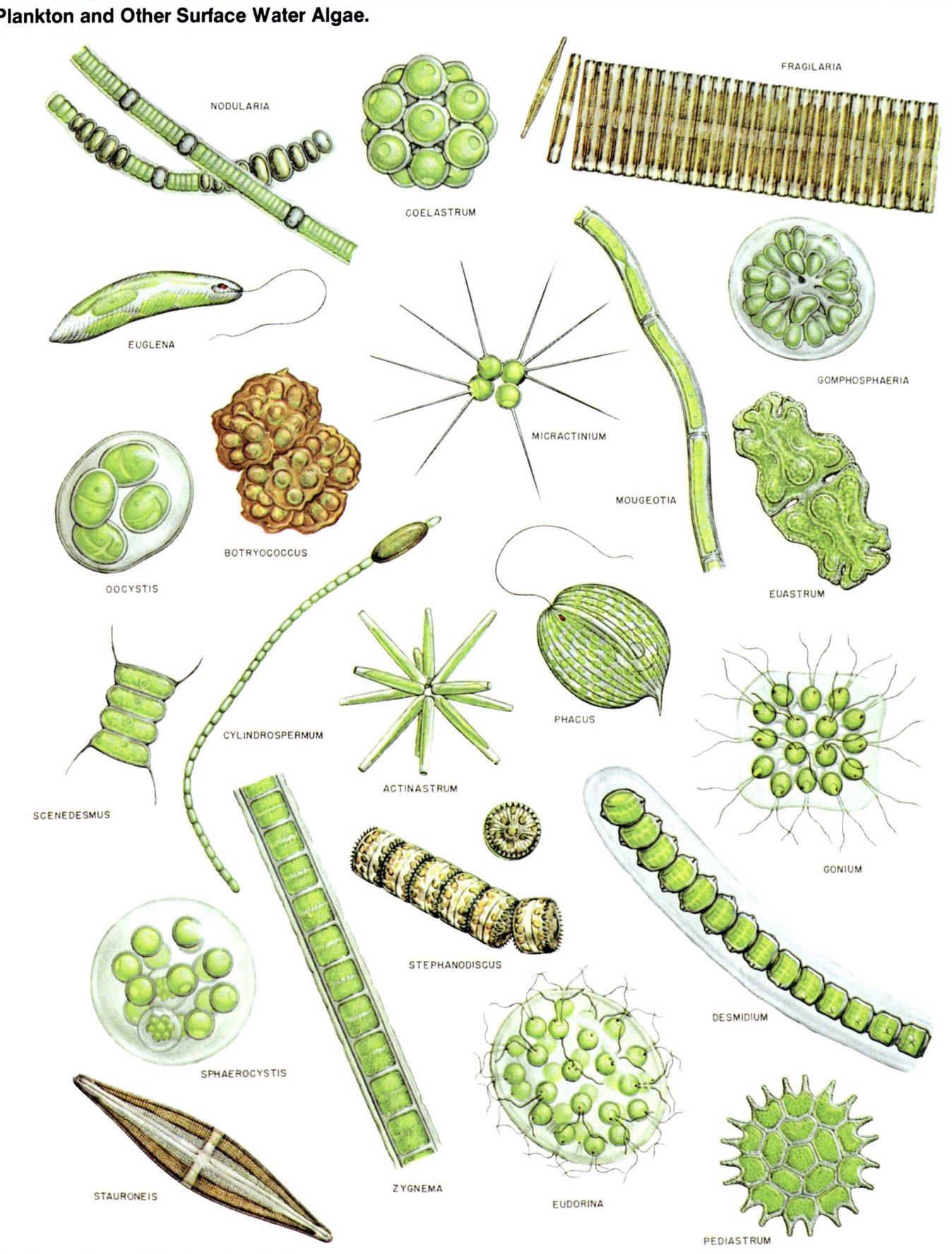

Algae Growing on Reservoir Walls.

Figure B-6

Figure B-6a

Estuarine Pollution Algae.

Figure B-6b

Sewage Pond Algae.

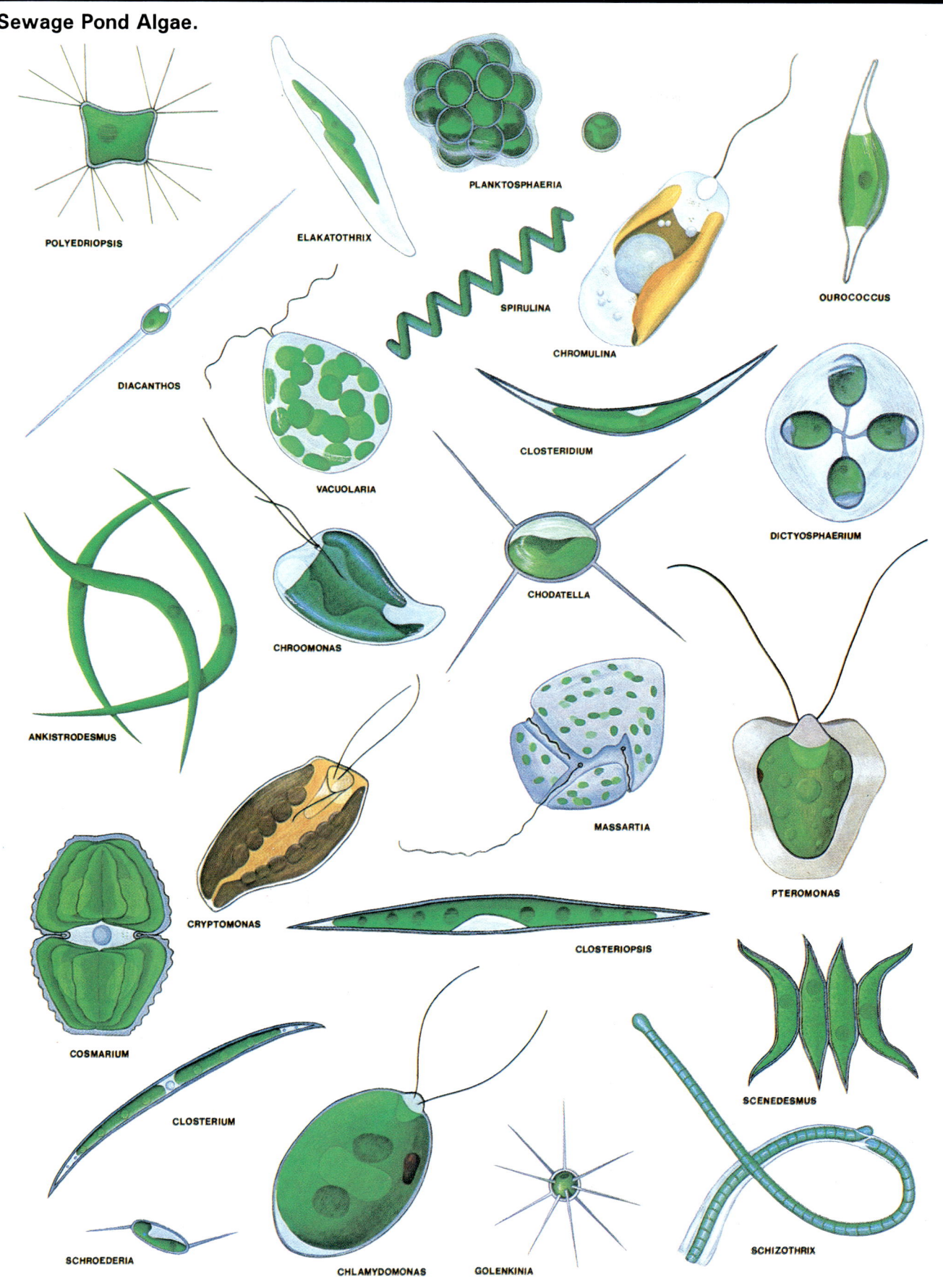

Figure B-7

Types of Freshwater Algae.

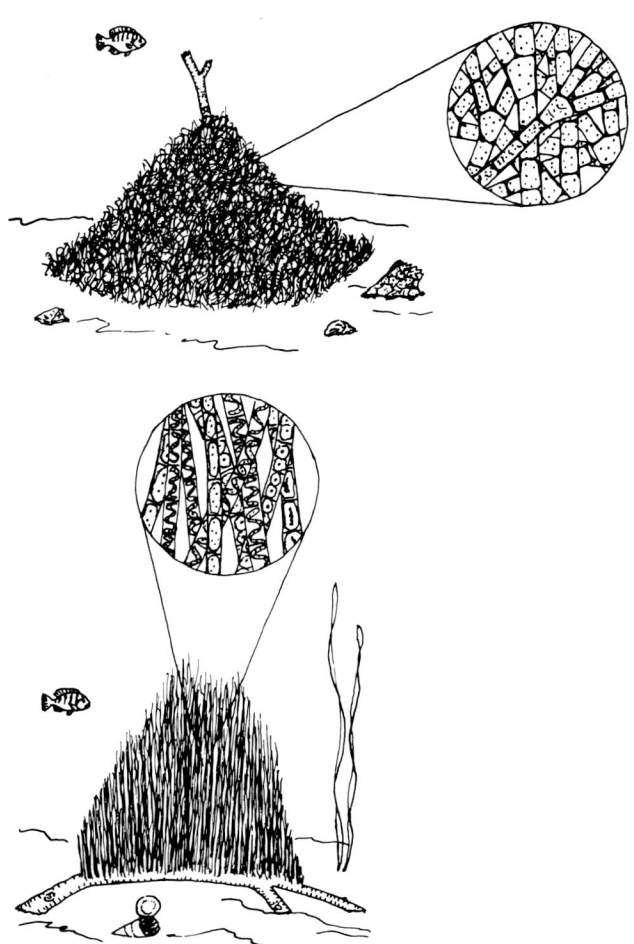

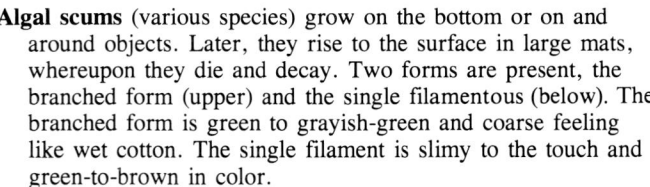

Algal scums (various species) grow on the bottom or on and around objects. Later, they rise to the surface in large mats, whereupon they die and decay. Two forms are present, the branched form (upper) and the single filamentous (below). The branched form is green to grayish-green and coarse feeling like wet cotton. The single filament is slimy to the touch and green-to-brown in color.

Waterlily (*Nymphaea*)
The large, circular, waxy floating leaves are deeply notched and borne on tough, elastic stems. The large white, pink, yellow, or blue flowers, with 12–40 petals, float on the surface with the leaves. The thick inter-twining roots of this plant form extensive mats over the bottom.

Table B–6b. — Sewage pond algae

Species Names	Linear Magnifications
Ankistrodesmus falcatus	1000
Chlamydomonas pertusa	2000
Chodatella quadriseta	3000
Chromulina vagans	2000
Chroomonas caudata	2500
Closteriopsis brevicula	250
Closterium acutum	500
Cosmarium botrytis	500
Cryptomonas cylindrica	2000
Diacanthos belenophorus	750
Dictyosphaerium ehrenbergianum	1000
Elakatothrix gelatinosa	1000

Table B–6b. — Sewage pond algae

Species Names	Linear Magnifications
Golenkinia radiata	500
Massartia vorticella	1500
Ourococcus bicaudatus	1500
Planktosphaeria gelatinosa	500
Polyedriopsis spinulosa	500
Pteromonas angulosa	1250
Scenedesmus dimorphus	1500
Schizothrix calcicola	2000
Schoederia setigera	250
Spirulina subtilissima	4000
Vacuolaria novo-munda	1250

Figure B-7

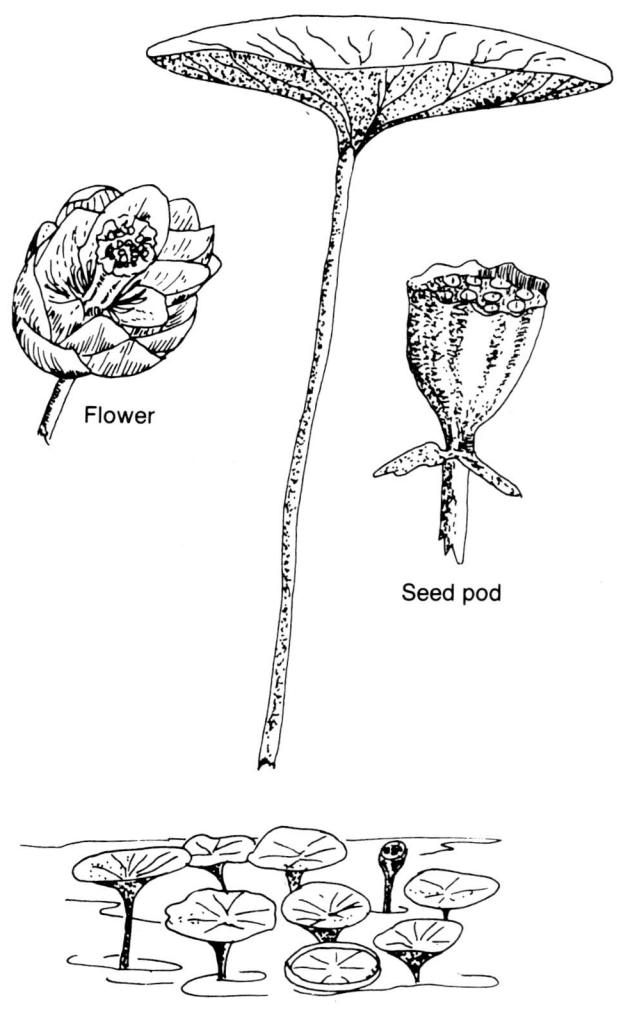
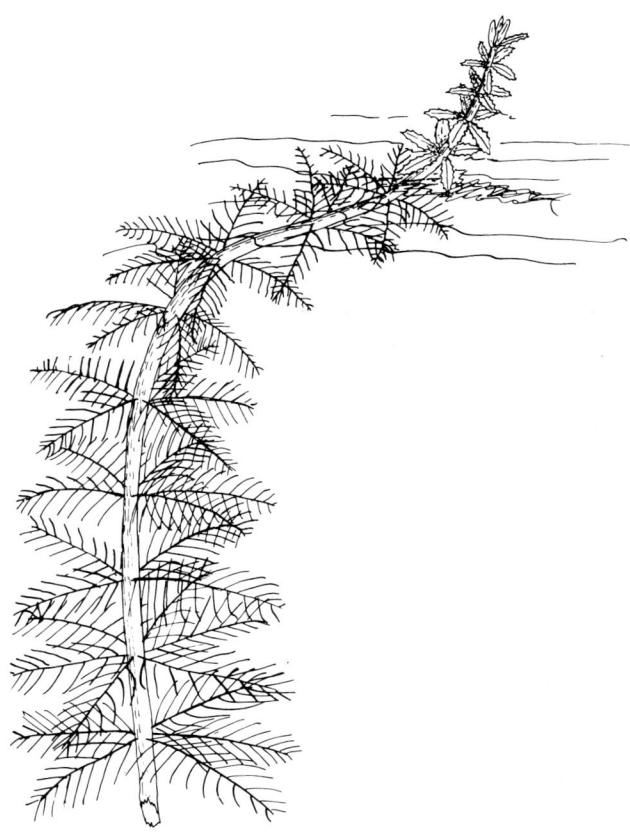

Lotus

Leaves. Circular 12–24 in. in diameter, with the centers "cupped." Usually they stand up out of the water, but immature leaves lie flat on the surface.

Stem. 1/4 to 1/2 in. in diameter, stiff and upright.

Flowers. 4.5 to 10 in. in diameter, pale yellow in color.

Special characteristics. The large, cupped leaves which stand upright are distinctive and characteristic of no other native plant.

Watermilfoil (*Myriophyllum*)

Leaves. Upper aerial ones elliptical with scalloped edges, giving it a prickly appearance, and dark green in color. Late in summer they turn red. Submerged leaves much longer and wider. Finely divided, giving the leaves a feather-like appearance.

Stems. Thick, reddish to brown, hollow or loosely pith filled.

Special characteristics. The upper leaves of this plant projecting 3–5 inches above the surface of the pond make it easily identified and separated from Parrots Feather.

Habitat. Shallow water 0–5 ft. deep.

 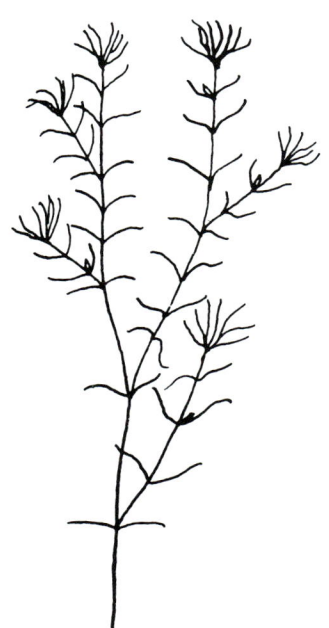 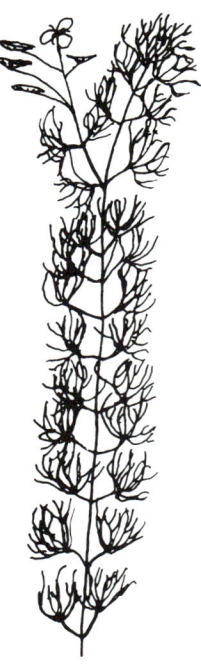

Coontail (*Ceratophyllum*)
A submerged, brittle herb with leaves in whorls about the main stem, which is generally forked once to several times. The leaves are very fine and forked (sometimes divided into threes) at the tips. These "tiplets" have a "spiny" appearance because of their wavy margins. This plant is "rooted" in the spring and early summer and free-floating in the late summer and early fall. The seeds of this plant are taken by waterfowl.

Naiads (*Najas*)
Submerged herbs with opposite or whorled, narrow to thread-like leaves. The bases of the leaves sheathe the stem. The main stem is branched and has fibrous roots. The seeds are small and elliptical and are found in the axil of the leaves. This plant is a favored food of many ducks.

Fanwort (*Cabomba*)
Delicate, branched, submerged herbs with finely divided leaves that are opposite or in whorls. Occasionally, the upper floating leaves are produced. These are small, oblong, and attached at the center of the blade. The flowers are small and have three white to yellow petals.

Figure B-7

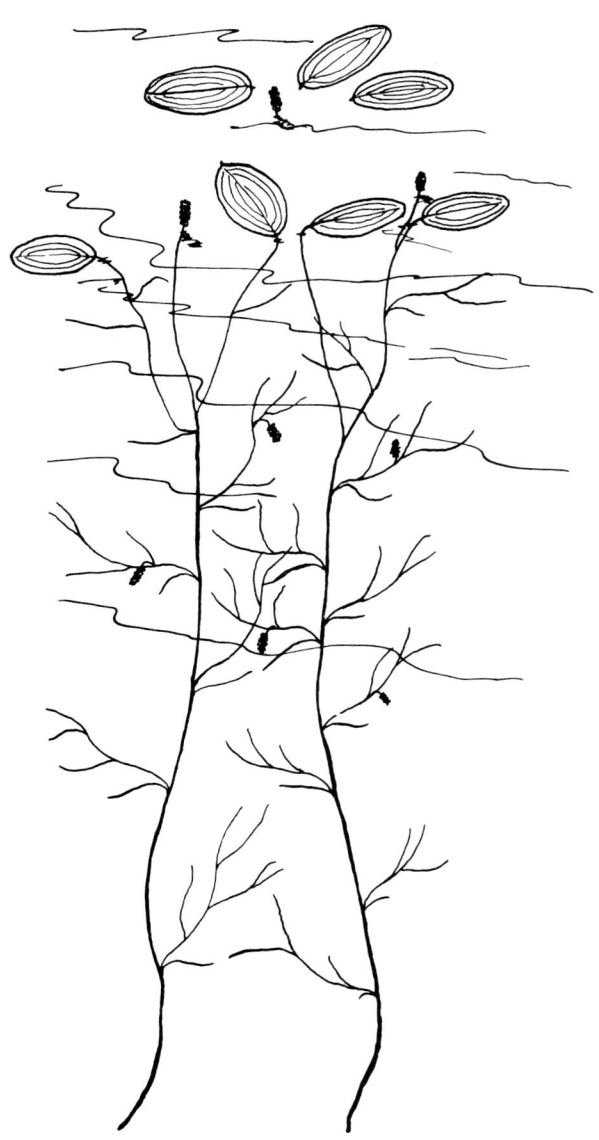

Pond Weeds (*Potamogeton*)

Leaves. Upper floating ones elliptical to oval, generally small (1/2-2 in). One species has very large leaves (3-10 in long). The surface is waxy. In some cases, the upper leaves may be missing. Lower leaves are very narrow, 1-2 mm (0.04-0.08 in) or less and strap-like.

Stems. Thin, but strong, varying in length, according to water depth. Always rooted.

Fruit. The small, cylindrical seed heads are on separate stalks, sometimes appearing above the water or generally found in the axil of the leaves. These seeds are avidly taken by ducks.

Special characteristics. The only plant having leaves this small floating on the surface of the water.

Habitat. Any body of water.

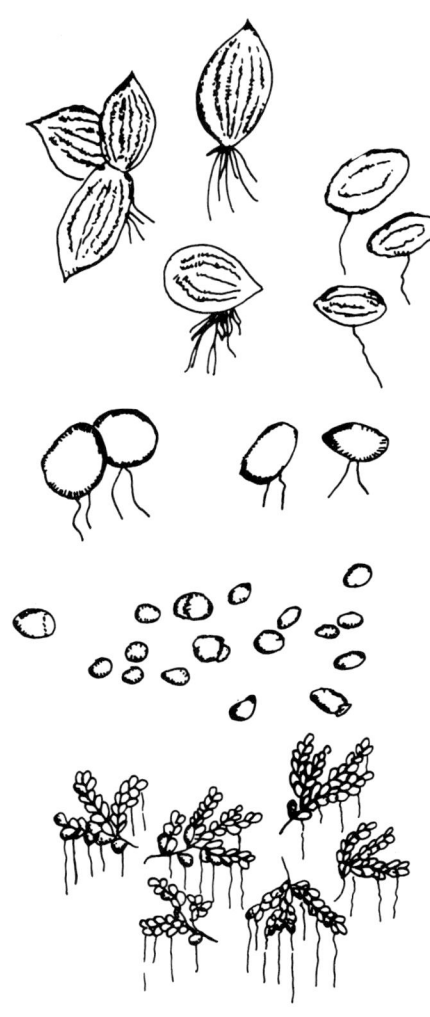

Free Floating Plants

Duckweeds, (*Lemna* and *Spirodela* spp.)
Small, 1-12 mm (0.04-4.7 in) long, free floating green plants of various shapes, generally oblong, that have one to many small rootlets hanging in the water and 1-15 "nerves" appearing on the top of the plants. *Spirodela* is larger and may be purple on the underside. These plants are 1-8 mm (0.04-0.3 in) long, oval in outline with a small pointed tip, have many rootlets, and have 5-11 "nerves" on the upperside. *Lemna* has 1-5 "nerves," one rootlet, and is green on the underside and smaller, being 1-5 mm (0.04-0.2 in). Under some conditions, these plants may have a reddish color. It would be best to check them to prevent confusion with water fern.

Duckmeal, (*Wolffia* and *Wolfella* spp.)
The tinest flowering plant in the world, 0.5-2mm (0.04-0.08 in) long, rootless, globular to ellipsoid in outline.

Waterfern (*Azolla sp.*)
Larger than the above, 0.5-1 cm (0.2-0.4 in) long, having small overlapping leaves borne on a once-to-several times forking stem. Several small roots hang in the water. At maturity these plants are red, rosy pink, or reddish brown. Young plants are green.

Figure B-7

Continued.

Elodea, or Waterweed (*Anacharis*)

Leaves. Narrow, gradually tapering to the tip. Borne either opposite each other in pairs, or in whorls of 4–5. Leaves of *E. densa* are large and coarse; those of the other two species smaller and more delicate.

Stems. Herbaceous, lax, and generally rooted; sometimes form floating mats.

Flowers. Arise between the stem and leaves. Three petals are present, and these are white or pinkish. Mainly spreads vegetatively.

Special characteristics. These are the same plants sold in pet stores for use in aquariums. Perennial, does not die back in the winter.

Habitat. Shallow water of lakes or ponds.

Watershield (*Brasenia*)

The floating leaves are oval and the undersides reddish and covered with a shiny covering. The stems are usually covered with this coating also. The small flowers are reddish to purple and have 3–4 petals. Prefers ponds or slow-moving acid water with a sandy bottom. Some diving ducks readily take the seeds of this plant.

Figure B-7

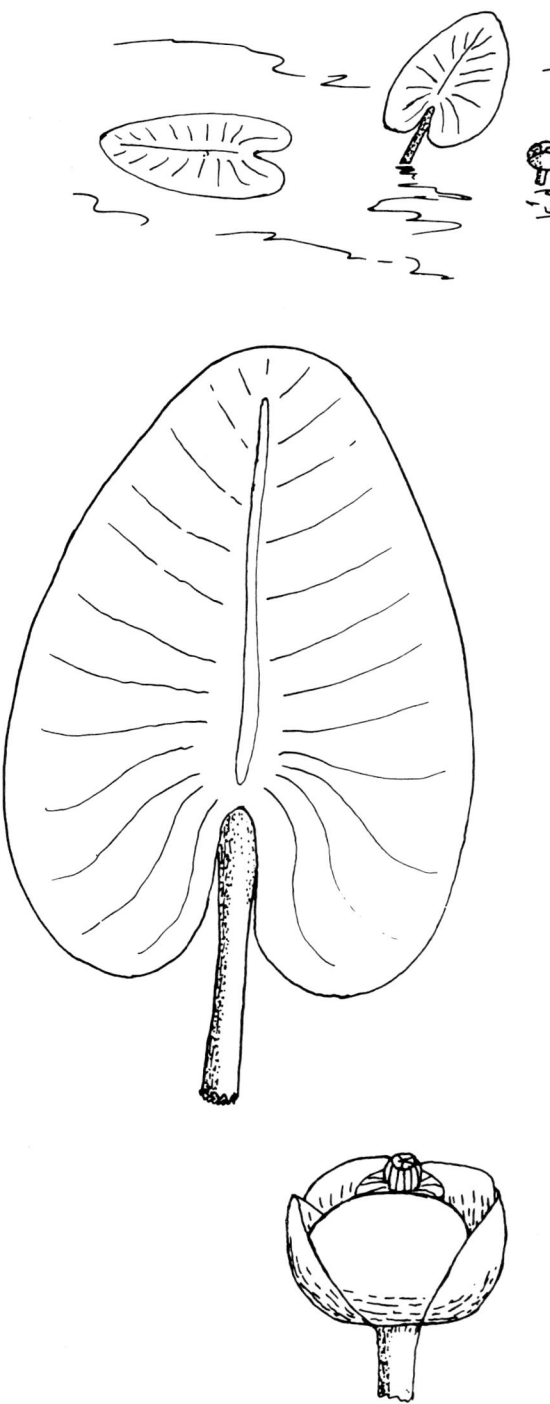

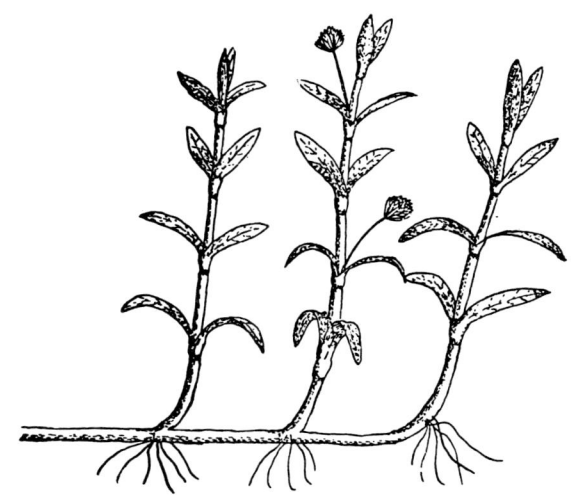

Alligatorweed (*Alternanthera*)
May be found growing upright on damp soil or growing as a floating mat in water. Leaves roughly oval and opposite one another on the stem. The bases of the leaves merge to form a sheath which is slightly swollen. Leaves and stems succulent and fleshy. Flowers white and resemble the flowers of white clover. These are borne on a long stalk growing between the stem and leaf. Seeds are not viable, and this plant reproduces vegetatively from the nodes.

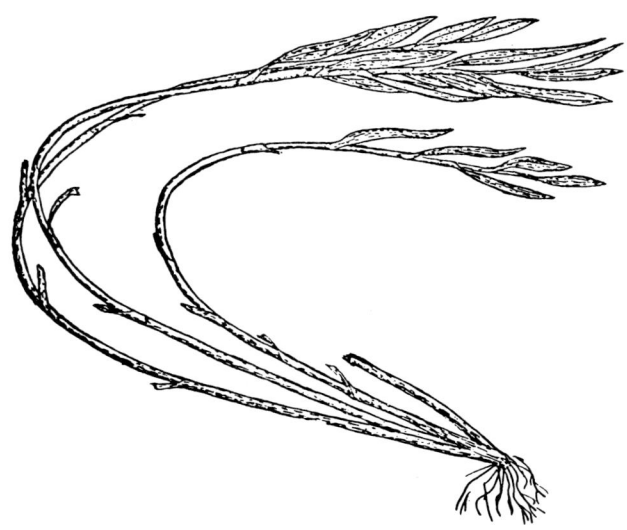

Spatterdock, Yellow Waterlily (*Nuphar*)
The large, waxy leaves are heart-shaped and may be upright or floating on the surface. The stems are thick, strong, and elastic. The small flowers are yellow and waxy in appearance. This plant is found in the shallows out to a depth of 3–4 feet.

Carolina Watergrass (*Hydrochloa*)
Leaves small (1–2 in long x 1/4 in wide), elliptical, grayish-green to green in color. These are found mainly towards the end of the stem and float on the surface. This plant can be found growing next to the shore or in shallow water. May form floating mats which can cover up small ponds. Rarely fruits.

Continued.

Smartweeds, Water Pepper (*Polygonum*) Plants inhabiting the shallow water of a pond, with lance-shaped, alternate leaves. At the base of each leaf is a sheath going around the stem and topped with long, fine hairs. The flowers are pink, white, or greenish and found in terminal spikes or on short, lateral spikes originating between the leaf and the stem. The seed is either triangular or lens-shaped in cross-section. These seeds are a choice food for ducks.

Lizardstail (*Saururus*)
Succulent herbs with jointed stems and alternate drooping heart-shaped leaves, found along the edges of the water. The long, nodding, white-flowered spike is present during the summer and easily distinguishes this plant.

Cattail (*Typha*)
Long, narrow, veinless, bluish-green leaves, sheathing at the base of the plant, and the familiar seed head are enough to identify this plant.

Arrow-arum *(Peltandra)* - Upper Left
The leaves are shaped like a barbed arrowhead and are borne on thick fleshy stems. The yellow "flowers" are enclosed in a green, partly opened, sac-like structure which terminates in a wrinkled tip. The skin of the fruit is green, purplish, or brown, and the seeds are enclosed in a gelatinous mass within the fruit.

Arrowhead (*Sagittaria*) - Upper Right
The leaves are highly variable, but are generally arrowhead-shaped, though the "barbs" may or may not be present, according to the species and water depth. The small white flowers are in whorls of three along the main stalk.

Pickerelweed (*Pontederia*) - Lower Left
The leaves are heart-shaped and are borne on thick stems. The flowers are bluish and found in a terminal spike.

Note: In the case where the flowers are absent, these plants may be differentiated by the veination of the leaves. See the illustration.

Figure B-7

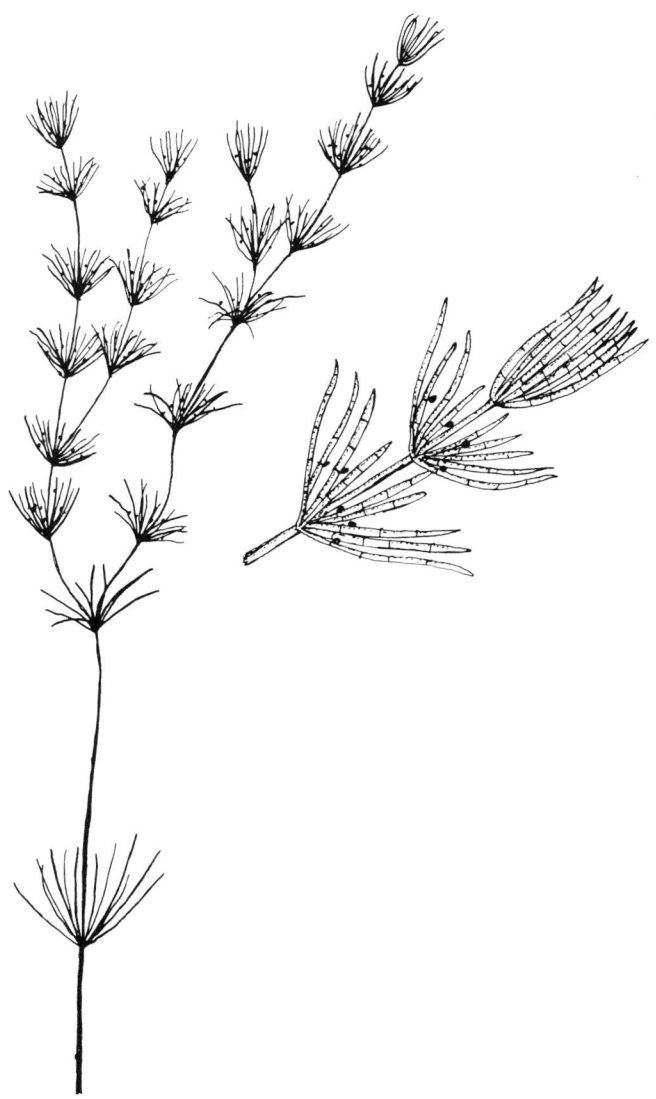

Stoneworts Muskgrass (*Chara*)
Actually a higher form of perennial algae. The 6–12 leaves are cylindrical and arranged in whorls around the stems and branches. Stems, branches, and leaves are very brittle, and when crushed, emit a strong musk-like or "skunky" odor. The "fruit" or *oogonium* appears as small black dots scattered over the leaves of the plant. Variable in height but generally not reaching the surface of the water. All parts of the plant are eaten by waterfowl.

Simple assessment of bottom-dwelling insects (ref. A-2).
Immature forms of bottom-dwelling stream insects live primarily in riffles—shallow, swift-flowing portions of a watercourse. Two major groups of aquatic insects should be present in the upper watercourse reaches of all unpolluted waterways: mayflies (fig. B-8) and caddisflies (fig. B-9). Mayflies have a roachlike body, a thin hairlike tail, and six jointed legs. Caddisflies have a maggotlike body, no tail, and six jointed legs.

To sample these insects, use the following simple technique. Remove three stones from a shallow, swift-flowing portion of the watercourse. Each stone should be about six inches in diameter. Place the stones in a bucket filled with stream water. Brush the entire surface of each stone with your hands. If after carefully examining the surface of each, you are satisfied that no insects remain, then discard the stone. Pour the contents of the bucket through a white handkerchief. Count the number of mayflies and caddisflies. Using the following illustrations, identify and count the number of insects belonging to the various groups.

If both mayflies and caddisflies are absent from the watercourse, then the watercourse is severely polluted. If only mayflies *or* caddisflies are present, then the watercourse is probably moderately polluted. If both mayflies and caddisflies are present, along with stoneflies (fig. B-10), then the stream is probably in good-to-excellent condition. Stoneflies resemble mayflies in having a roachlike body, tail, and six jointed legs. Mayfly legs come to a fine point at the tips, whereas stonefly legs are tipped with two hooks or claws.

Insect larvae, which are tolerant of pollution and might be found in either clean or moderately polluted water, are blackflies, bloodworm midges (chironomids), rat-tailed maggots, and others. See also fig. A-4.

Figure B-8

Mayflies (Ephemeroptea).

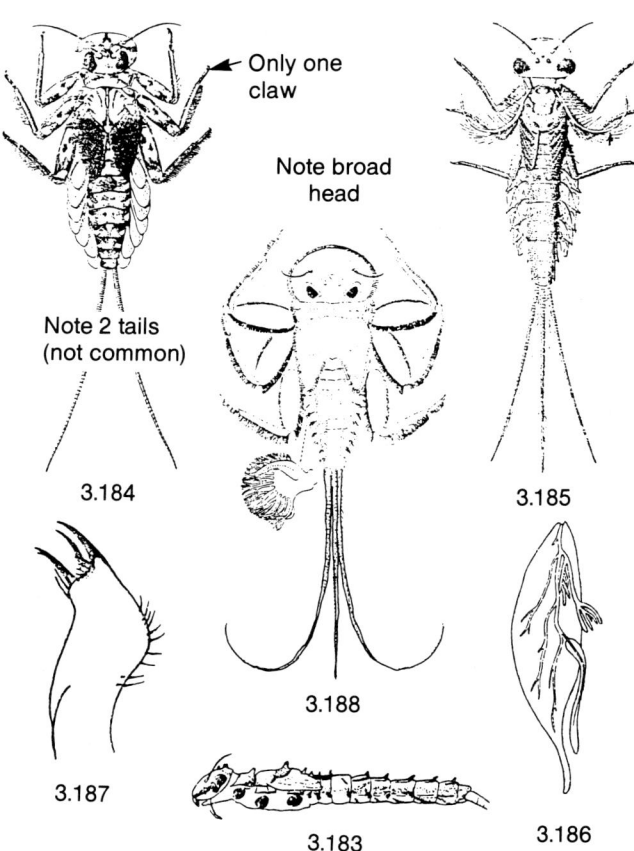

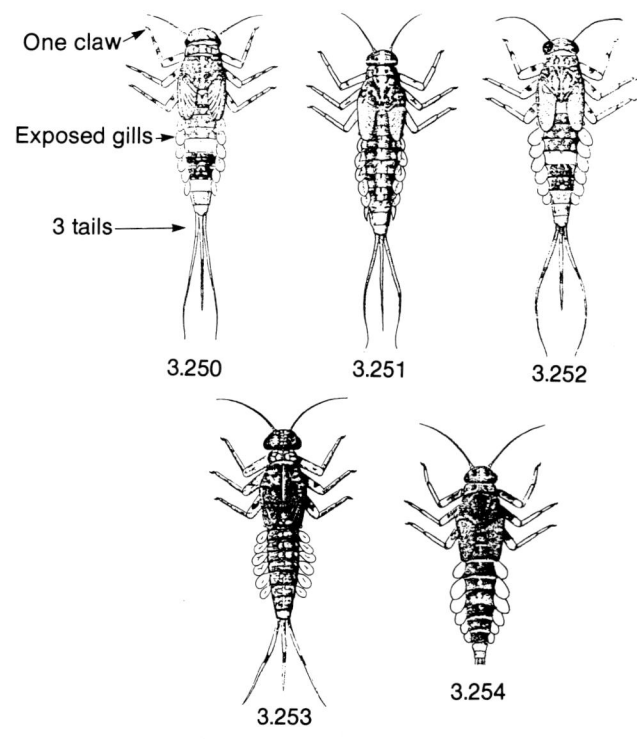

Figures 3.183 through 3.188. *Heptageniidae*. 3.183 *Spinadis*, lateral aspect (shown without legs and gills); nymph, dorsal aspect: 3.184 *Epeorus*, 3.185 *Arthroplea bipunctata*; 3.186 *Pseudiron*, gill of 3rd abdominal segment; *Anepeorus*: 3.187 mandible, 3.188 dorsal aspect and left abdominal gill (3.183 after Flowers and Hilsenhoff 1975; 3.184 through 3.186, 3.188, Illinois Natural History Survey, (INHS)).

Figures 3.250 through 3.254. *Baetidae. Baetis* nymph, dorsal aspect: 3.250 *B. macdunnoughi*, 3.251 *B. pygmaeus*, 3.252 *B. intercalaris*, 3.253 *B. propinquus*, 3.254 *B. tricaudatus* (all after Morihara and McCafferty, 1979).

Figure B-9

Caddisflies *(Tricoptera).*

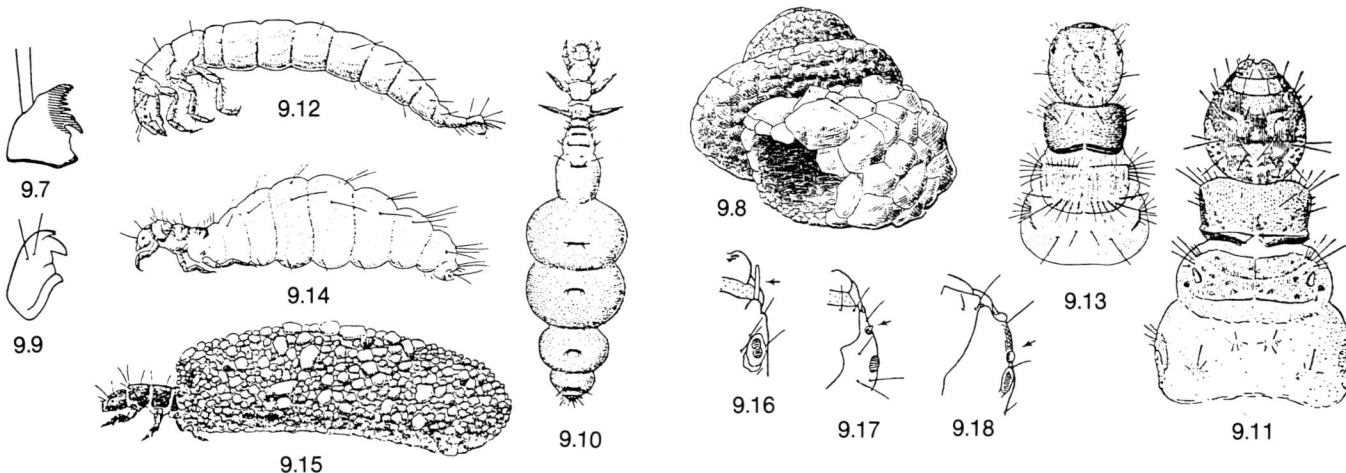

Figures 9.7 through 9.18. Family characters. *Helicopsyche borealis* larva: 9.7 anal claw, lateral aspect, 9.8 larval case; 9.9 *Brachycentrus*, anal claw of larva, lateral aspect; 9.10 *Leucotrichia pictipes* larva, dorsal aspect; 9.11 *Limnephilus submonilifer,* head and thorax of larva, dorsal aspect; 9.12 *Polycentropus* larva, lateral aspect; 9.13 *Lepidostoma*, head and thorax of larva, dorsal aspect; 9.14 *Hydroptila* larva, lateral aspect; 9.15 *Ochrotrichia* larva in purse case, lateral aspect; larval head, dorsal aspect: 9.16 *Leptocerus americanus*, 9.17 *Limnephilus*, 9.18 *Lepidostoma* (all INHS).

Figure B-10

Stoneflies *(Plecoptera).*

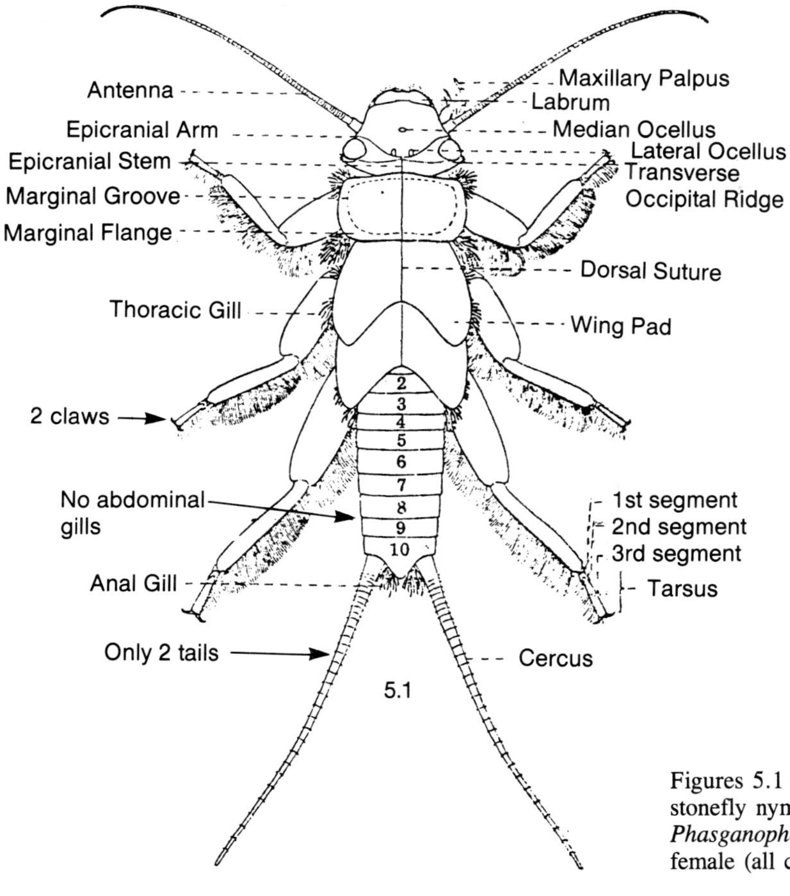

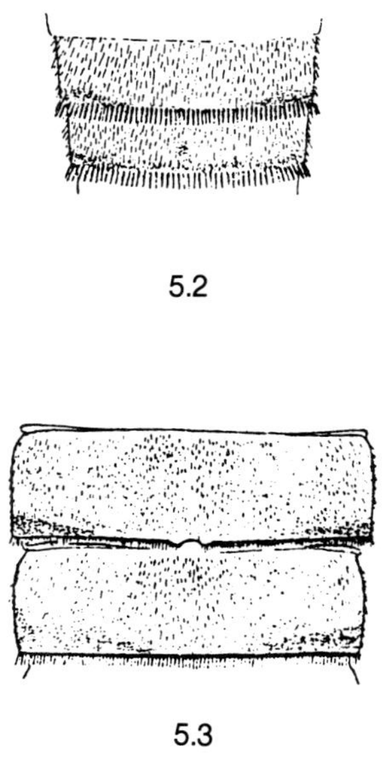

Figures 5.1 through 5.3. Stonefly structure. 5.1 generalized stonefly nymph, dorsal aspect. Abdominal segments 8 and 9 of *Phasganophora capitata* nymphs, ventral aspect: 5.2 male, 5.3 female (all courtesy of the INHS).

Figure B-10

Stoneflies, continued.

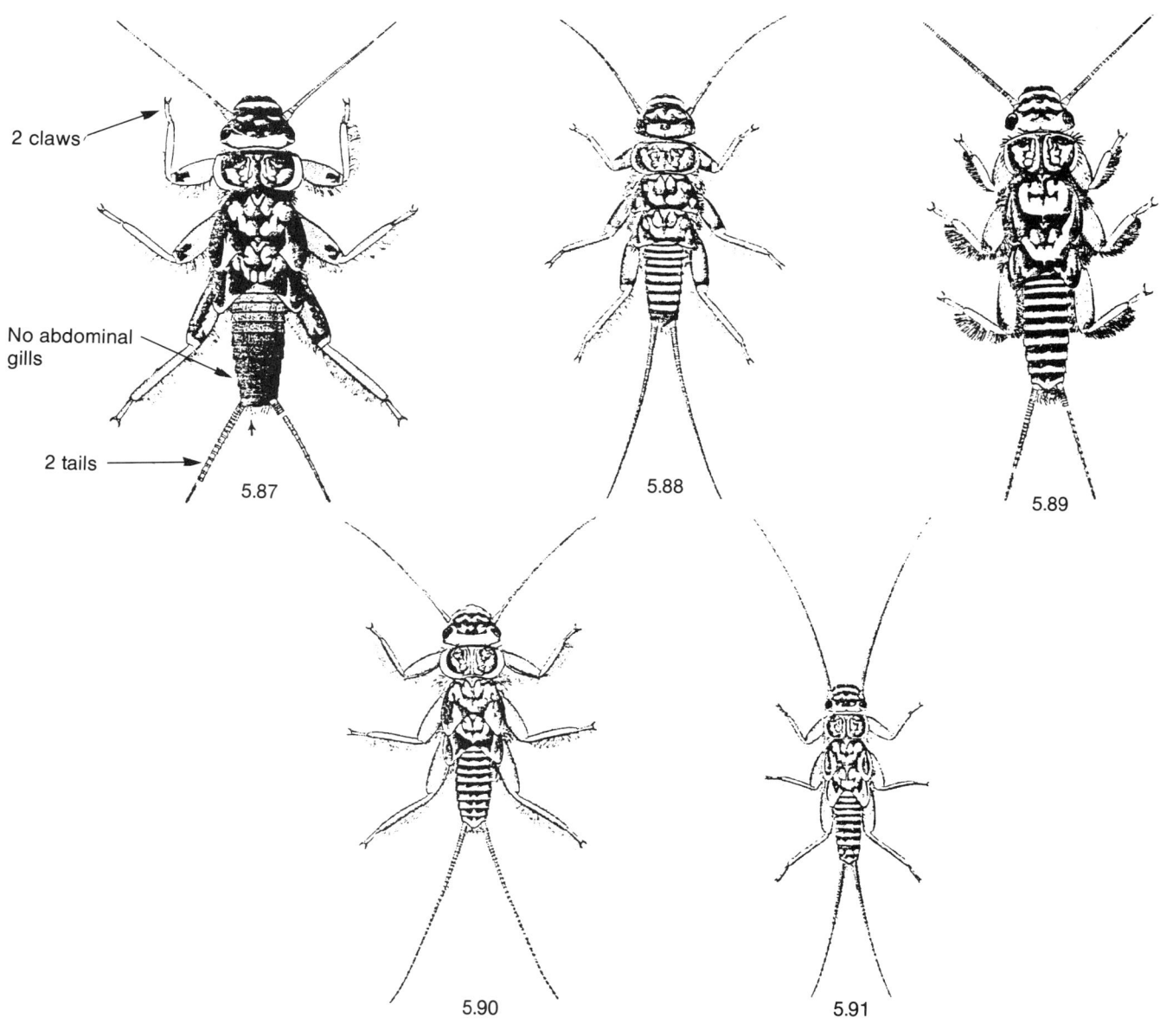

Figures 5.87 through 5.91. *Perlidae. Acroneuria* nymph, dorsal aspect: 5.87 *A. mela,* 5.88 *A. lycorias,* 5.89 *A. evoluta,* 5.980 *A. perplexa,* 5.91 *A. filicis* (all INHS).

Figure B-11

Key to the Major Invertebrate Species of Stream Zones.

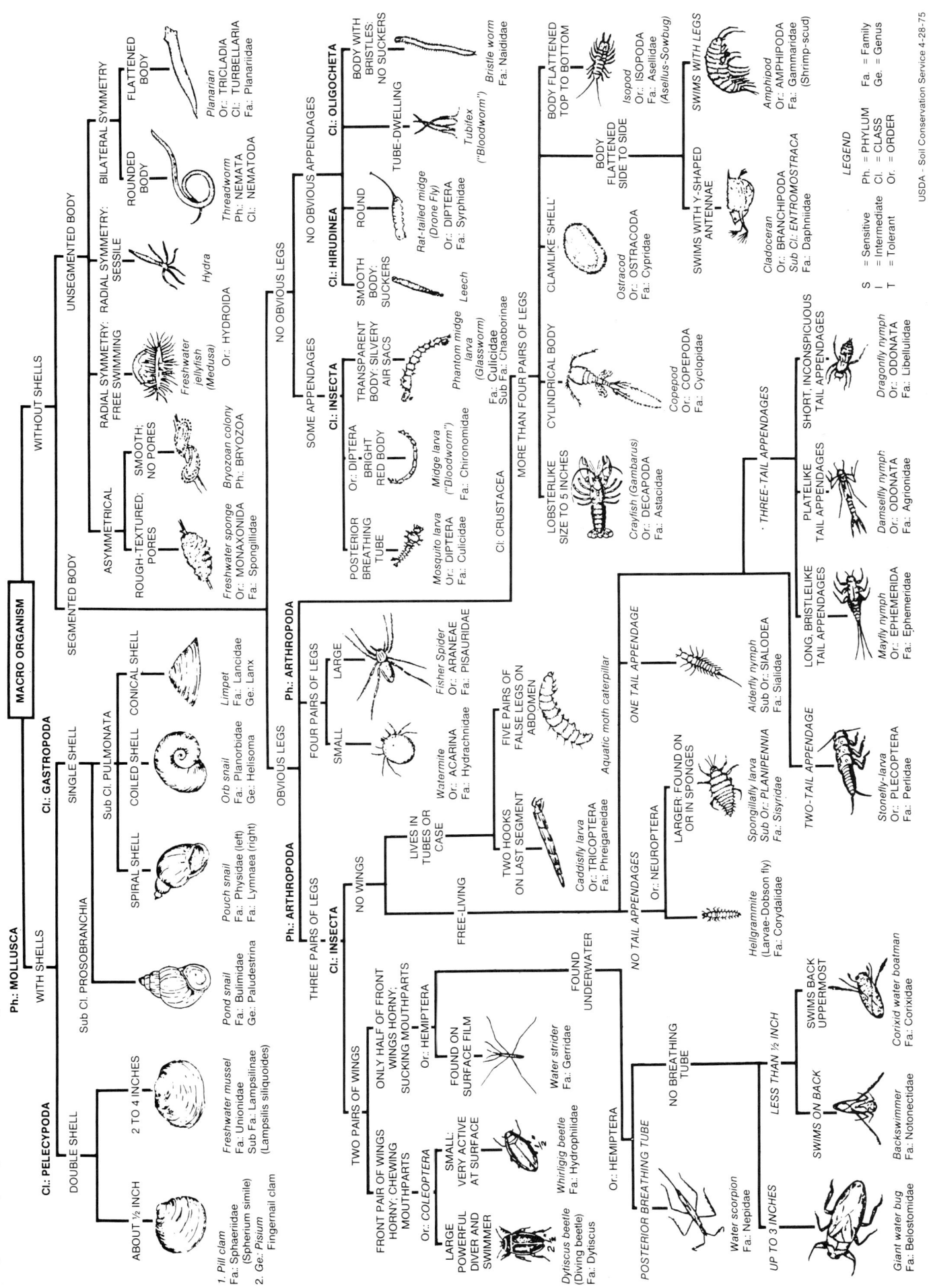

Figure B-12

Some Common Freshwater Fishes (Ref. B-4).

Alosa aestivalis (Mitchill)
Blueback herring

Order Clupeiformes
Family Clupeidae

TYPE LOCALITY: New York (Mitchill 1815. Trans. Lit. Philos. Soc. N.Y. 1:355-492).

SYSTEMATICS: Formerly placed in *Pomolobus,* synonymized most recently under *Alosa* by Svetovidov (1964. Copeia: 118-30). Often confused with *A. pseudoharengus.*

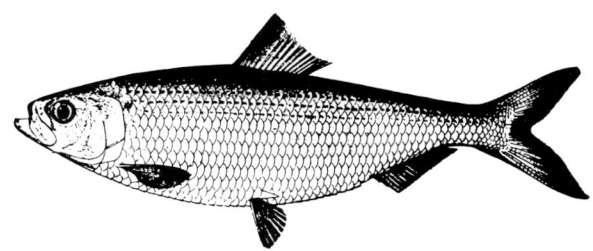

Washington, D.C., market ca. 23 cm SL (Jordan and Evermann 1900).

Alosa pseudoharengus (Wilson)
Alewife

Order Clupeiformes
Family Clupeidae

TYPE LOCALITY: Probably Delaware River at Philadelphia, Philadelphia Co., PA (Wilson ca. 1811 *in Rees' New Cyclopedia* 9: no pagination).

SYSTEMATICS: Formerly placed in *Pomolobus,* most recently synonymized with *Alosa* (Svetovidov 1964. Copeia: 118-30). Often confused with *A. aestivalis.*

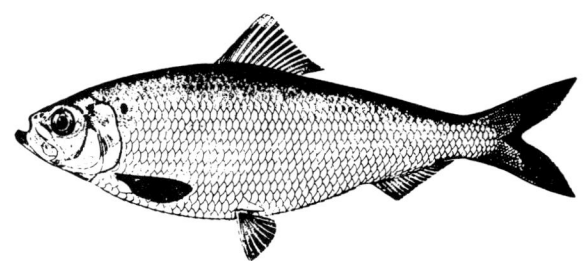

Washington, D.C., market ca. 26 cm SL (Jordan and Evermann 1900).

Alosa sapidissima (Wilson)
American shad

Order Clupeiformes
Family Clupeidae

TYPE LOCALITY: Probably Delaware River at Philadelphia, Philadelphia Co., PA (Wilson ca. 1811. *in Rees' New Cyclopedia* 9: no pagination).

SYSTEMATICS: Forms geographically disjunct species pair with *A. alabamae* (Berry 1964. Copeia: 720-30). Meristic differences seen between spawning populations inhabiting various river systems (and their tributaries) along Atlantic coast (Carscadden and Leggett 1975. J. Fish. Res. Board Can. 32:653-60 and included references).

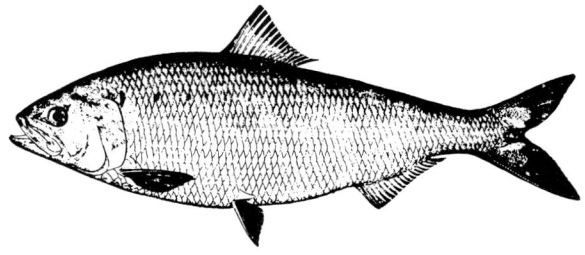

VA: Norfolk, ca. 43 cm SL (Jordan and Evermann 1900).

Figure B-12

Continued.

Oncorhynchus gorbuscha (Walbaum)
Pink salmon

Order Salmoniformes
Family Salmonidae

TYPE LOCALITY: Rivers of Kamchatka, USSR (Walbaum *in* Artedi 1772. *Genera Piscium* 3:4-723).

SYSTEMATICS: Essentially unstudied, apart from Rounsefell's (1962. Fishery Bull. 62:237-70) work on relationships between *Oncorhynchus* species. Vladykov (1962. Bull. Fish. Res. Board Can. 136:1-172) compared pyloric caeca in specimens from North America and Japan. Taxonomic comparisons between even and odd year stocks seem warranted.

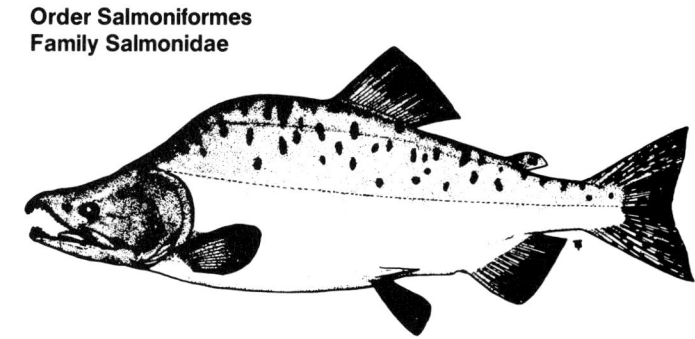

ca. 53 cm SL (NMC).

Oncorhynchus tshawytscha (Walbaum)
Chinook salmon

Order Salmoniformes
Family Salmonidae

TYPE LOCALITY: Rivers of Kamchatka, USSR (Walbaum *in* Artedi 1792. *Genera Piscium* 3:4-723).

SYSTEMMATICS: Broad meristic variation within species, but individual stocks usually uniform. Scott and Crossman (1973. *Freshwater Fishes of Canada)* provided comparison of variation between Pacific and introduced Lake Ontario populations.

CA: Sacramento Co., American River, male, 64 cm SL (Moyle 1976).

Salmo gairdneri Richardson
Rainbow trout

Order Salmoniformes
Family Salmonidae

TYPE LOCALITY: Mouth of Columbia River at Fort Vancouver, WA (Richardson 1836. *Fauna Boreali-Americana*).

SYSTEMATICS: The "rainbow trout" is comprised of two major groups, coastal rainbow trouts and redband trouts. The redband trout, native to headwaters of McCloud River, CA, is closely related to the golden trout of Kern River drainage, CA, *S. aguabonita.* Oldest name for any member of redband trout group is *S. newberryi.* Oldest name applied to any member of either group is *S. mykiss,* proposed by Walbaum in 1792 for the Kamchtakan trout. Many practical difficulties are involved if *gairdneri* becomes synonym of *mykiss.*

(N.C. Wildl. Resour. Comm. and NCSM)

Figure B-12

Salmo trutta Linneaus
Brown trout

Order Salmoniformes
Family Salmonidae   EXOTIC

TYPE LOCALITY: "Europe" (Linneaus 1758. *Systema naturae,* Laurentii Salvii, Holmiae, 10th ed., 1:1-824).

SYSTEMATICS: Subgenus *Salmo.* Rather variable within native range and number of subspecies recognized. Hybridizes with *Salvelinus fontinalis* in nature (hybrids called "tiger trout") and artificially hybridized with other salmonids (Scott and Crossman 1973. *Freshwater Fishes of Canada;* Buss and Wright 1958. Trans. Am. Fish. Soc. [1957] 87:172-81).

(N.C. Wildl. Resour. Comm. and NCSM)

Cyprinus carpio Linnaeus
Common carp

Order Cypriniformes
Family Cyprinidae   EXOTIC

TYPE LOCALITY: Europe (Linneaus 1758. *Systema naturae,* Laurentii Salvii, Holmiae 10th ed., 1:1-824).

SYSTEMATICS: Subfamily Cyprininae, which does not include native North American cyprinids. Hybridizes with goldfish, *Carassius auratus,* another exotic cyprinine (Scott and Crossman 1973. *Freshwater Fishes of Canada).* Hubbs *(in* Blair [ed.] 1961. *Vertebrate Speciation: A Symposium.)* discussed Asiatic genus *Carassiops,* possibly of ancient hybrid origin between *C. carpio* and *C. auratus.*

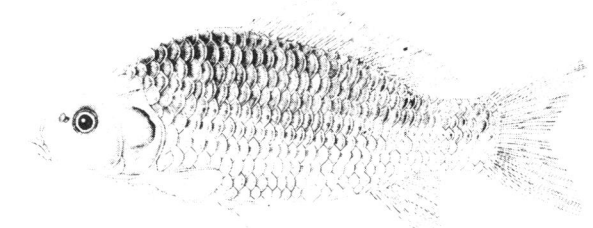

MD: Charles Co., Community Lake, 151 mm SL (NCSM).

Notemigonus crysoleucas (Mitchill)
Golden shiner

Order Cypriniformes
Family Cyprinidae

TYPE LOCALITY: New York (Mitchill 1814. Rept. on Fishes of New York: 1-30).

SYSTEMATICS: Possibly more closely related to certain Eurasian cyprinids than to any North American group (Gosline 1974. Jap. J. Ichthyol. 21: 9-15). Three subspecies have been recognized—*N. c. crysoleucas* in northeast, and *N. c. auratus* and *N. c. bosci* in south—but recent authors have not considered these valid. Variation in anal fin ray count appears to be influenced by water temperature during development (Hubbs 1921. Trans. Ill. State Acad. Sci. 11: 147-51; Schultz 1927. Pap. Mich. Acad. Sci. Arts Letts. [1926] 7:417-32). Scott and Crossman (1973. *Freshwater Fishes of Canada)* discussed and provided additional data on geographic variation in this character.

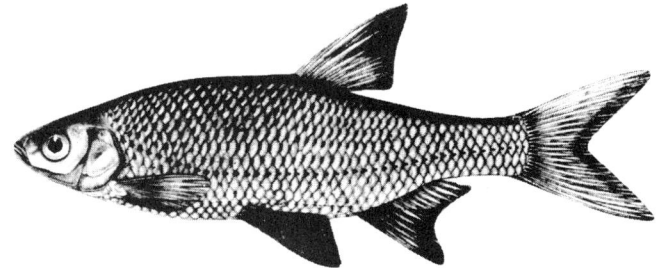

MD: Anne Arundel Co., Lake Waterford, 101 mm SL (NCSM).

Figure B-12

Continued.

Notropis cornutus (Mitchill)
Common shiner

Order Cypriniformes
Family Cyprinidae

TYPE LOCALITY: Wallkill River, 4.8 km sw of New Paltz, Ulster Co., NY (Mitchill 1817. Am. Mon. Mag. Crit. Rev. 1:289-90).

SYSTEMATICS: Subgenus *Luxilus*, Gilbert (1964. Bull. Fla. State Mus. Biol. Sci. 8:95-194) reviewed systematics of species. Hybridizes extensively with *N. chrysocephalus* (Gilbert 1961. Copeia:181-92). Based on blood protein patterns Menzel (1976. Biochem. Syst. Ecol. 4:281-93) considered *N. cornutus* and *N. chrysocephalus* as subspecies. *N. albeolus* is also closely related to *N. cornutus* and replaces it on middle Atlantic coast.

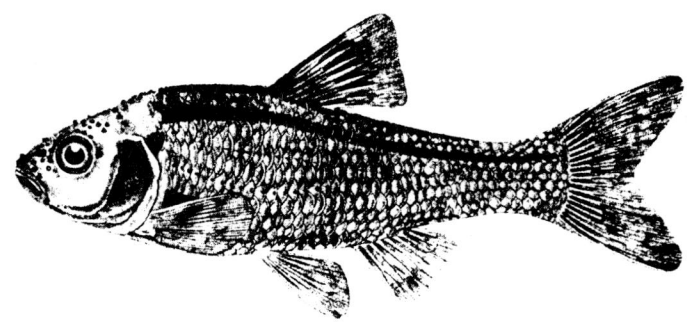

MD: Harford Co., Swan Creek, male, 88 mm SL (NCSM).

Rhinichthys atratulus (Hermann)
Blacknose dace

Order Cyprinformes
Family Cyprinidae

TYPE LOCALITY: "North America" (Hermann 1804. Observationes Zoologicae, quibus novae complures, aliaeque anamalium species descibuntur et illustrantur 31:1-332).

SYSTEMATICS: Three subspecies distributed about as follows: *R. a. atratulus* on Atlantic slope; *R. a. meleagris* in central and northern interior; and *R. a. obtusus* (including nominal form *simus*) from lower Ohio basin to upper Mobile drainage (Hubbs 1936. Copeia: 124-25; Matthews et al. ms). Matthews et al. (1979. Abstr. 59th Ann. ASIH meetings) discussed intergradation between *R. a. atratulus* and *R. a. obtusus* in James River drainage, VA.

MD: Charles Co., Zekiah Swamp, 51 mm SL (NCSM).

Catostomus commersoni (Lacepede)
White sucker

Order Cypriniformes
Family Catostomidae

TYPE LOCALITY: None given (Lacepede 1803. *Histoire Naturelle Poissons* 5:1-803).

SYSTEMATICS: No comprehensive analysis of systematics over entire range published, although numerous dwarf populations have received individual recognition (McPhail and Lindsey 1970. *Freshwater Fishes of Northwestern Canada and Alaska*). Beamish and Crossman (1971. J. Fish. Res. Board Can. 34:371-78) concluded dwarf form *C. commersonii utawana* not valid subspecies. Metcalf (1966. Univ. Kans. Publ. Mus. Nat. Hist. 17:23-189) suggested that three geographical forms from eastern, Plains, and Hudson Bay drainages existed in past.

MD: Frederick Co., Glade Creek, 96 mm SL (NCSM).

Figure B-12

Ictalurus catus (Linnaeus)
White catfish

Order Siluriformes
Family Ictaluridae

TYPE LOCALITY: "Northern part of America" (Linnaeus 1758. *Systema naturae* Laurentii Salvii, Holmiae, 10 ed., 1:1-824).

SYSTEMATICS: No definitive study; no subspecies recognized. Phylogenetic relationships to other ictalurids presented by Taylor (1969. U.S. Natl. Mus. Bull. 282:1-315).

CA: Lake Co., Clear Lake, 11 cm SL (Moyle 1976).

Ictalurus melas (Rafinesque)
Black bullhead

Order Siluriformes
Family Ictaluridae

TYPE LOCALITY: "Ohio River" (Rafinesque 1820. Q. J. Sci. Lit. Arts 9:48-55).

SYSTEMATICS: Two subspecies sometimes recognized: *I. Melas catulus* from Gulf coast states and northern Mexico, and *I. m. melas* from farther north (Smith 1979. *The Fishes of Illinois;* Scott and Crossman 1973. *Freshwater Fishes of Canada)*. List of synonyms provided by Scott and Crossman (1973). Phylogenetic relationships with other ictalurids presented by Taylor (1969. U.S. Natl. Mus. Bull. 282:1-315), and Lundberg (1975. Univ. Mich. Mus. Zool. Pap. Paleo. 11).

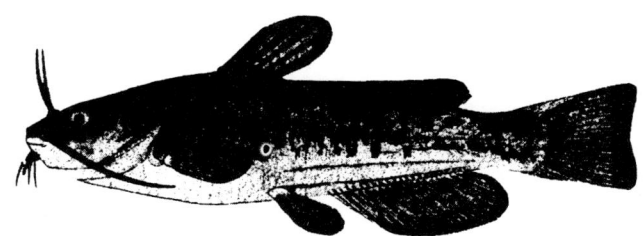

MD: Anne Arundel Co., Annapolis Reservoir, 99 mm SL (NCSM).

Ictalurus punctatus (Rafinesque)
Channel catfish

Order Siluriformes
Family Ictaluridae

TYPE LOCALITY: "Ohio River" (Rafinesque 1818. Am. Mon. Mag. Crit. Rev. 3:354-56).

SYSTEMATICS: Bailey et al. (1954. Proc. Acad. Nat. Sci. Phila. 106:109-64) discussed geographic and clinical variation but did not recognize subspecies. Possibly name-worthy forms were originally present, but situation has become greatly (perhaps hopelessly) confused by extensive introductions within and outside original range. Several closely related Mexican species, but precise relationships yet to be delineated. Most closely related United States species is *I. lupus* of TX and Mexico. Phylogenetic relationship to other ictalurids presented by Taylor (1969. U.S. Nat. Mus. Bull. 282:1-315).

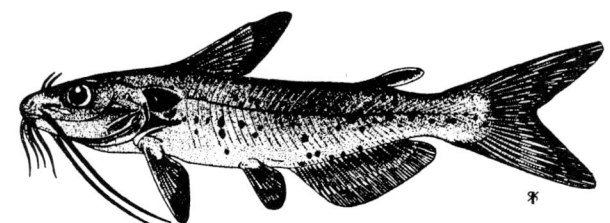

MD: Cecil Co., Susquehanna River, 127 mm SL (NCSM).

Figure B-12 Continued.

Noturus gyrinus (Mitchill)
Tadpole madtom

TYPE LOCALITY: Wallkill River, NY (Mitchill 1817. Am. Monthly Mag. Crit. Rev. 1:289-90).

SYSTEMATICS: Subgenus *Schilbeodes*. Appears to be most closely related to *N. lachneri* (Taylor 1969. U.S. Natl. Mus. Bull. 282:1-315).

Order Siluriformes
Family Ictaluridae

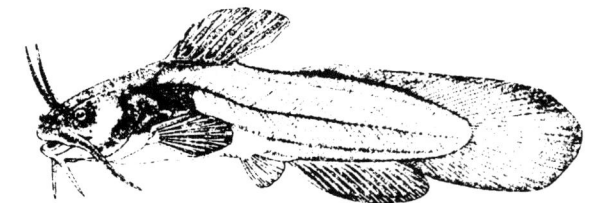

MD: St. Mary's Co., St. Mary's River (NCSM)

Morone saxatilis (Walbaum)
Striped bass

TYPE LOCALITY: "New York" (Walbaum *in* Artedi 1792. *Genera Piscium* 3:4-723).

SYSTEMATICS: Appears in earlier literature as *Roccus lineatus*. Whitehead and Wheeler (1966. Ann. Mus. Civ. Stor. Nat. Genova 76:23-41) showed that *Morone* has priority over *Roccus*.

Order Perciformes
Family Percichthyidae

(N.C. Wildl. Resour. Comm. and NCSM)

Lepomis macrochirus Rafinesque
Bluegill

TYPE LOCALITY: "Ohio River" (Rafinesque 1819. J. Physique 88:417-29).

SYSTEMATICS: Three subspecies are recognized. *Lepomis m. macrochirus* occurs in the Great Lakes and north Mississippi basin, *L.m. speciosus* in TX and Mexico and *L.m. purpurescens* on the Atlantic slope from coastal VA to FL (Hubbs and Lagler 1964. *Fishes of the Great Lakes Region*). Widespread introductions have resulted in extensive mixing of these gene pools. Avise and Smith (1974. Evolution 28:42-56) studied geographic variation and subspecific intergradation, and Avise and Smith (1977. Syst. Zool. 26:319-35) studied relationships to other centrarchid species using electrophoretic data. Commonly hybridizes with several other species of *Lepomis*, particularly in areas of ecological disturbance. Considered to be most closely related to *L. humilis* (Branson and Moore 1962. Copeia:1-108).

Order Perciformes
Family Centrarchidae

(N.C. Wildl. Resour. Comm. and NCSM)

Figure B-12

Micropterus salmoides (Lacepede)
Largemouth bass

Order Perciformes
Family Centrarchidae

TYPE LOCALITY: "les rivieras de le carolina"; Charleston, SC, regarded as probable type locality (Lacepede 1802. *Histoire Naturelle des Poissons* 4:1-728).

SYSTEMATICS: Subfamily Lepominae, tribe Micropterini. Formerly placed in monotypic genus *Huro* (Hubbs 1926. Misc. Publ. Mus. Zool. Univ. Mich. 15:1-77; Hubbs and Bailey 1940. Misc. Publ. Mus. Zool. Univ. Mich. 48:1-51). Hubbs and Bailey (1940) reviewed systematics, and Bailey and Hubbs (1949. Occas Pap. Mus. Zool. Univ. Mich. 516:1-40) defined and mapped distinctive subspecies, *M. s. floridanus*, endemic to peninsular FL.

(N.C. Wildl. Resour. Comm. and NCSM)

Pomoxis annularis Rafinesque
White crappie

Order Perciformes
Family Centrarchidae

TYPE LOCALITY: "Ohio River" (Rafinesque 1818. Am. Mon. Mag. Crit. Rev. 4:39-42).

SYSTEMATICS: Subfamily Centrarchinae, tribe Centrarchini. Branson and Moore (1962. Copeia: 1-108) studied morphology of acoustico-lateralis system and determined closest generic relationships to be with *Centrarchus*. Avise et al. (1977. Copeia: 250-58), based on electrophoretic data, sugested relationships might be closer to *Lepomis* and *Micropterus*, subfamily Lepominae. Bailey (1938. Ph.D. diss., Univ. Michigan) reviewed systematics. Known to hybridize naturally with *P. nigromaculatus*; artificially crossed with other genera (Schwartz 1972. Publ. Gulf Coast Res. Lab. Mus. 3:1-328).

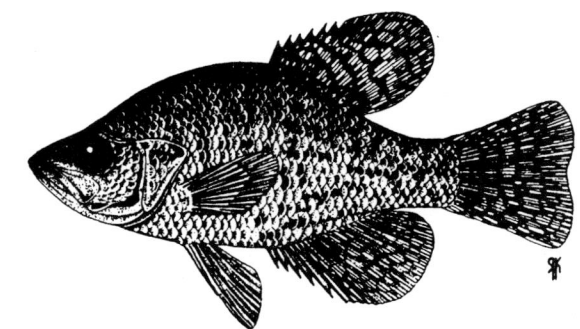

MD: Garrett Co., Piney Creek, 165 mm SL (NCSM).

Lepomis megalotis (Rafinesque)
Longear sunfish

Order Perciformes
Family Centrarchidae

TYPE LOCALITY: Kentucky, Licking, and Sandy rivers, KY (Rafinesque 1820. *Ichthyologia Ohiensis*).

SYSTEMATICS: Closest relative *L. marginatus*, these two species comprising subgenus *Icthelis*. Hybridizes extensively with other *Lepomis*. Most polytypic member of family Centrarchidae, consisting of from four to six subspecies. Presently under study by compiler.

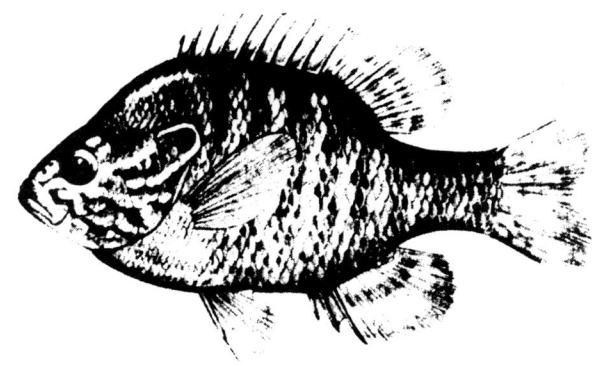

(NMC)

Figure B-12

Continued.

Lepomis microlophus (Günther)
Redear sunfish

Order Perciformes
Family Centrarchidae

TYPE LOCALITY: St. Johns River, FL (Günther 1859. *Catalogue of the Fishes in the British Museum* 1:1-524).

SYSTEMATICS: Bailey (1938. Ph.D. diss., Univ. Michigan) concluded that *L. microlophus* comprises two distinct subspecies. Extensive introductions of stocks into ranges of each other have obscured natural relationships. Avise and Smith (1977. Syst. Zool. 26:319-35) on basis of electrophoretic data determined that most closely related species of *Lepomis* likely are *L. megalotis* and *L. marginatus*.

(N.C. Wildl. Resour. Comm. and NCSM)

Micropterus dolomieui Lacepede
Smallmouth bass

Order Perciformes
Family Centrarchidae

TYPE LOCALITY: None given (Lacepede 1802. *Histoire Naturelle des Poissons* 4:1-728).

SYSTEMATICS: Hubbs and Bailey (1940. Misc. Publ. Mus. Zool. Univ. Mich. 48:1-51) recognized two subspecies: *M. d. dolomieui* east of Mississipi River and from central MO north; and *M. d. velox* from middle Arkansas River drainage. Intergrades identified from White and Black river drainages, AR and MO, and Ouachita River system, AR. Widely introduced and genetic integrity of original stocks may no longer be valid. Summary of nonmenclature in Scott and Crossman (1973. *Freshwater Fishes of Canada*).

(N.C. Wildl. Resour. Comm. and NCSM)

Pomoxis nigromaculatus (Lesueur)
Black crappie

Order Perciformes
Family Centrarchidae

TYPE LOCALITY: Wabash River, OH (Lesueur *in* Cuvier and Valenciennes 1829. *Histoire Naturelle des Poissons* 3:1-500).

SYSTEMATICS: Subfamily Centrarchinae, tribe Centrarchini. Branson and Moore (1962. Copeia:1-108) studied morphology of acoustico-lateralis system and determined closest generic relationships to be with *Centrarchus*. Avise et al. (1977. Copeia: 250-58), based on electrophoretic data, suggested relationships might be closer to *Lepomis* and *Micropterus*, subfamily Lepominae. Bailey (1938. Ph.D. diss., Univ. Michigan) reviewed systematics. Known to hybridize naturally with *P. annularis;* artificially crossed with other genera (Schwartz 1972. Publ. Gulf Coast Res. Lab. Mus. 3:1-328).

(N.C. Wildl. Resour. Comm. and NCSM)

***Cottus bairdi* Girard**
Mottled sculpin

Order Perciformes
Family Cottidae

TYPE LOCALITY: Mahoning River, OH (Girard 1850. Proc. Am. Assoc. Adv. Sci. [1849]:409-11).

SYSTEMATICS: Bailey and Bond (1963. Occas. Pap. Mus. Zool. 634:1-27) presented summary of species included in *C. bairdi* group. Considerable geographic variation throughout wide range of species, and overall systematic picture unresolved. Some populations classified as *C. bairdi* may be distinct species. Scott and Crossman (1973. *Freshwater Fishes of Canada*) noted that Canadian populations have received insufficient attention for subspecific recognition. Robins (1954. Ph.D. diss., Cornell Univ.) studied systematics in eastern United States. McAllister (1964. J. Fish. Res. Board Can. 21:1339-42) discussed separation of *C. bairdi* from *C. cognatus*.

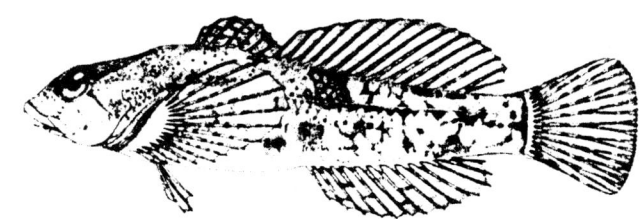

(NCSM)

Detection of *Escherichia coli* in water samples. The presence of *E. coli* is detected by the following procedure:

A water sample is collected in a sterile bottle and poured into a filtering apparatus. When water is drawn through a sterile filter, the bacterial contaminants are left behind on a piece of filter paper. This filter paper is placed in a sterile Petri plate containing a nutrient broth, which the bacteria will use to grow. The plate is incubated at 35 degrees C for 24 hours. Portable incubators are available that run off a car's cigarette lighter and can be used until a source of electricity is available. By the end of this 24-hour period, individual *E. coli* organisms have divided to produce metallic-green colonies visible to the naked eye. The log, or geometric mean of 200 fecal coliform colonies per five 100 ml samples collected over 30 days, is the allowable limit for fresh waters used for swimming. *See Standard Methods for the Examination of Water and Wastewater* (1985) for details (ref. B-5).

APPENDIX C

Glossary

Acute toxicity. A relatively short-term lethal or other adverse effect to a test organism caused by pollutants, and usually defined as occurring within 4 days for fish and large invertebrates, and shorter times for smaller organisms.

Alluvial soil. A deposit of sand, mud, etc., formed by flowing water.

Animal waste. Either solid or liquid products, resulting from digestive or excretory processes, and eliminated from an animal's body.

Aquifer. Any geological formation containing water, especially one that supplies water for wells, springs, etc.

Bedrock. Unbroken solid rock, overlain in most places by soil or rock fragments.

Best management practice. An engineered structure or management activity, or combination of these, that eliminates or reduces an adverse environmental effect of a pollutant.

Bioaccumulation. The process of a chemical accumulating in a biological food chain by being passed from one organism to another as the contaminated organism is preyed upon by another organism.

Biochemical oxygen demand (BOD). An empirical test in which standardized laboratory procedures measure the oxygen required for the biochemical degradation of organic material, and the oxygen used to oxidize inorganic materials, such as sulfides and ferrous iron.

Biomass. The total weight of all living organisms or of a designated group of organisms in a given area.

Birth defect. A deformity of an organism at birth that results from a biologic infection, genetic anomaly, or presence of a pollutant during the gestation period.

Chronic toxicity. A relatively long-term adverse effect to a test organism caused by or related to appetite changes, growth, metabolism, reproduction, a pollutant, genetic mutation, etc.

Cobble streambed. A watercourse predominately lined with naturally rounded stones, rounded by the water's action. Size varies from a hen's egg to that used as paving stones.

Conservation practice. An engineered structure or management activity that eliminates or reduces an adverse environmental effect of a pollutant and conserves soil, water, plant, or animal resources.

Confined aquifer. An aquifer bounded above and below by impermeable beds of rock or soil strata or by beds of distinctly lower permeability than that of the aquifer itself.

Cultural eutrophication. The process whereby human activities increase the amounts of nutrients entering surface waters, giving increased algal and other aquatic plant population growths, resulting in accelerated eutrophication of the watercourse or water body.

Delta. A nearly flat, often triangular, plain of deposited sand, mud, etc., between diverging branches of a river mouth.

Dissolved oxygen (DO). The amount of oxygen dissolved in water. Generally, proportionately higher amounts of oxygen can be dissolved in colder waters than in warmer waters.

Emergent rooted plant. An aquatic plant whose roots are in the watercourse or water body's bottom and whose upper part emerges from or lies on top of the water.

Ephemeral stream. A watercourse that flows briefly only in direct response to precipitation in the immediate locality, and whose channel is at all times above the water table.

Escherichia coli (E. coli). A bacterium of the intestines of warm-blooded organisms, including humans, that is used as an indicator of water pollution for disease-producing organisms.

Eutrophication. A natural process whereby a watercourse or water body receives nutrients and becomes more biologically productive, possibly leading to a water body clogged with aquatic vegetation.

Feathering. The process whereby dissolved salts move upward through a wooden post or stake and become deposited on the structure's outer surface, yielding a white, fluffy, "feathery" appearance.

Fertilizer. Any substance used to make soil or water more productive. Fertilizers may be commercially produced or be the result of animal or plant activities.

Food chain. The transfer of food energy from plants through a series of organisms by repeated eating and being eaten.

Food web. An interlocking pattern of several to many food chains.

Herbaceous vegetation. Plants having a stem that remains soft and succulent during the growing season, not woody.

Herbicide. A type of pesticide, either a substance or biological agent, used to kill plants, especially weeds.

Insecticide. A type of pesticide, either a substance or biological agent, used to kill insects or insect-like organisms.

Intermittent stream. A watercourse that flows only at certain times of the year, receiving water from springs or surface sources; also, a watercourse that does not flow continuously, when water losses from evaporation or seepage exceed available stream flow.

Invertebrate. An organism without a backbone.

Karst topography. An area of limestone formations characterized by sinks, ravines, and underground streams. Areas with less than 20 feet of soil over fractured limestone. No shale layers present, capping the top aquifer, but shale layers can separate the top aquifer from deeper ones.

Lake. A body of fresh or salt water of considerable size, whose open-water and deep-bottom zones (no light penetration to bottom) are large compared to the shallow-water (shoreline) zone, which has light penetration to its bottom.

Lentic water. Water that is standing, not flowing, such as that in a lake, pond, swamp, or bog.

Lotic water. Water that is flowing or running, such as that in a spring, stream, or river.

Macrophyte. Any large plant that can be seen without the aid of a microscope or magnifying device. Examples of aquatic macrophytes are cattail, bulrush, arrowhead, waterlily, etc.

Mancos shale. A geologic formation, remnant of an ancient sea, which exists in many parts of the western United States. When irrigation waters flow through the formation, salts become dissolved in the water, increasing its salinity.

Mesotrophic water body. A water body classified midway between oligotrophic and eutrophic; characterized by moderate amounts of nutrients entering the water body, a moderate number of shoreline aquatic plants, and occasional plankton blooms.

Methemoglobinemia. The presence of methemoglobin in the blood, making the blood useless as a carrier of oxygen. Methemoglobin, a compound closely related to oxyhemoglobin, is found in the blood following poisoning by certain substances, such as nitrate. Young babies, both human and animal, are particularly susceptible to methemoglobinemia, leading to a condition known as "blue baby," which if untreated can cause death.

Mudcap. A thick deposit of mud or fine sediment lying over permeable materials.

Mud plastering. Mud deposited by force of water against the sides of a watercourse, sealing them.

Nonpathogenic organism. An organism that does not produce disease.

Nonpoint source pollution. "Diffuse" pollution, generated from large areas with no particular point of pollutant origin, but rather from many individual places. Urban and agricultural areas generate nonpoint source pollutants.

Nontarget organisms. Plants or animals that inadvertently are sprayed by pesticide when "target" vegetation or animals are missed by the spraying operation.

Nutrient. Any substance, such as fertilizer phosphorous and nitrogen compounds, which enhances the growth of plants and animals.

Oligotrophic water body. A water body characterized by few nutrients entering the water body, few to no shoreline aquatic plants, and rarely any plankton blooms.

Overland flow. Water flow over the land, often in "sheet" flow or in small rivulets before emptying into a defined watercourse.

Pathogenic organism. An organism that produces disease.

Periphyton. Small-to-microscopic aquatic plants, which grow on stones, submerged twigs, and other plants. Their appearance may be that of a coating on these objects.

Perennial stream. Watercourse that flows continuously throughout the year and whose upper surface generally stands lower than the water table in the area adjacent to the watercourse.

Pesticide. Any chemical or biological agent that kills plant or animal pests. Herbicides, insecticides, nematocides, miticides, algicides, etc., are all pesticides.

Photosynthesis. The process by which plants manufacture their own food (simple carbohydrates) from carbon dioxide (CO_2) and water. The plant's chlorophyll-containing cells use light as an energy source and release oxygen as a byproduct.

Phytoplankton. Small-to-microscopic, aquatic, floating plants.

Piping. Under low dissolved oxygen conditions, the act of fish coming to surface of the water and capturing a bubble of air in their mouth.

Plankton. Small-to-microscopic, floating or feebly swimming, aquatic plants and animals.

Point source pollution. Pollutants originating from a "point" source, such as a pipe, vent, or culvert.

Pond. A body of fresh or salt water, smaller than a lake, and where the shallow-water zone (light penetration to its bottom) is relatively large compared to the open water and deep bottom (no light penetration to the bottom).

Pool. In a watercourse, an area often following a rapids (riffle), which is relatively deep with slowly moving water compared to the rapids.

Protected bedrock. Areas with 50 feet or more of fine-to-medium textured soils and a shale layer capping the topmost bedrock aquifer.

Receiving waters. Waters of a watercourse or water body that receive waters from overland flow or other watercourses.

Resource Management System (RMS). A combination of conservation practices and management identified by the primary use of land or water. Under an RMS, the resource base is protected by meeting acceptable soil losses, maintaining acceptable water quality, and maintaining acceptable ecological and management levels for the selected resource use.

Riffle. In a watercourse, an area often upstream of a pool, which is relatively shallow with swiftly moving water compared to the pool.

Riparian zone. An area, adjacent to and along a watercourse, which is often vegetated and constitutes a buffer zone between the nearby lands and the watercourse.

Runoff. Water that runs off the land in sheet flow, in rivulets, or in defined watercourses.

Runoff curve number. An index number, used to approximate the amount of runoff resulting from a given rainfall event.

Saline seep. Water, carrying salts, rising to the surface usually in a localized area, after traveling subterraneously from another location. Saline seep salts can reduce productivity or kill plants, leaving a barren place in the field or landscape.

Scoliosis. A vertebral deformity, such as "broken back" syndrome in fish, resulting from a biologic infection, genetic anomaly, or the presence of a pollutant.

Shallow bedrock. Areas having 20 to 50 feet of soil capping the topmost bedrock aquifer. No shale layer present, capping the topmost aquifer, but shale layers may separate top aquifers from deeper ones.

Sinkhole. A circular depression, commonly funnel-shaped, in a Karst area. Drainage is subterranean; size is measured in meters or tens of meters.

Submergent rooted plant. An aquatic plant whose roots are in the watercourse or water body's bottom with the upper part of the plant submerged below the surface of the water. Pond weeds *(Potamogeton)* and muskgrass *(Chara)* are examples.

Teratology. The science or study of monstrosities or abnormal formations in animals or plants.

Turbidity. The presence of sediment in water, making it unclear, murky, or opaque.

Water body. An enlargement of a watercourse or a geologic basin filled with water, such as a lake or a pond.

Watercourse. A linear depression containing flowing water, such as a stream, creek, run, river, canal, ditch, etc.

Woody vegetation. Plants having a stem or trunk that is fibrous and rigid.

Zooplankton. Small-to-microscopic, aquatic, floating animals.

APPENDIX D

References

1-1. J. Ball. *Stream Classification Guidelines for Wisconsin.* Department of Natural Resources Technical Bulletin; Madison, WI, 1982.

1-2. *Report to Congress: Nonpoint Source Pollution in the U.S.* United States Environmental Protection Agency, Water Planning Div., Washington, D.C. 1984.

1-3. *The Second RCA Appraisal: Soil, Water, and Related Resources on Nonfederal Land in the United States—Analysis of Condition and Trends.* U.S. Dept. of Agriculture, Soil Conservation Service, Washington, D.C., 1987.

1-4. WATSTORE—The National Water Data Storage and Retrieval System. U.S. Geological Survey, Water Resources Div., Reston, VA.

1-5. *Basic Statistics—1982 National Resources Inventory.* U.S. Dept. of Agriculture, Soil Conservation Service, Statistical Bulletin No. 756, 1987.

1-6. Surface Soil Surveys. U.S. Geological Survey, Reston, VA.

1-7. Soil Survey Laboratory Data State Reports. U.S. Dept. of Agriculture, Soil Conservation Service, Soil Survey Div., Washington, D.C.

1-8. National Stream Quality Accounting Network (NASQAN). U.S. Geological Survey, Water Resources Div., Reston, VA.

3-1. J.M. Lawrence and L.W. Weldon. "Identification of Aquatic Weeds." *Hyacinth Control Jour.* (now *Jour. of Aquatic Plant Management)*, Vol. 4, pp. 5-17, 1965.

3-2. L.M. Cowardin, V. Carter, F.C. Golet, and E.T. LaRoe. *Classification of Wetlands and Deepwater Habitats of the United States.* Office of Biological Services, U.S. Fish and Wildlife Service, U.S. Dept. of the Interior, Washington, D.C., 1979.

3-3. E.L. Horwitz. *Our Nation's Lakes.* U.S. Environmental Protection Agency, Office of Water Regulations and Standards, Washington, D.C., (EPA 440/5-80-009), 1980.

3-4. D.D. Chiras. *Environmental Science: A Framework for Decision Making.* Benjamin/Cummings Pub. Co., Inc., Menlo Park, CA, 1985.

3-5. O.S. Owen. *Natural Resource Conservation: An Ecological Approach,* 4th Ed., Macmillan Pub. Co., New York, NY, 1985.

3-6. E.A. Keller. *Environmental Geology,* 3rd Ed., Charles E. Merrill Pub. Co., Columbus, OH, 1982.

3-7. M.P. Keown. *Streambank Protection Guidelines.* U.S. Army Corps of Engineers Waterways Experiment Station, Vicksburg, MS, 1983.

3-8. J.G. Needham and P.R. Needham. *A Guide to the Study of Fresh-water Biology.* Holden-Day, Inc., San Francisco, CA, 1962.

3-9. L.D. Marriage and R.F. Batchelor. "Ever-Changing Network of Our Streams, Rivers and Lakes." *Using Our Natural Resources,* Yearbook of Agriculture, U.S. Dept. of Agriculture, Washington, D.C., 1983.

3-10. W.M. Beck, Jr. "Studies in Stream Pollution Biology." *Quart. Jour. Fla. Acad. Sci.* 17(4):211-227, 1954.

3-11. K.F. Lagler. *Freshwater Fishery Biology.* 2nd Ed. Wm. C. Brown Co., Dubuque, IA, 1966.

3-12. G.W. Lewis. *Common Freshwater Sportfishes of the Southeast.* Cooperative Extension Service, U. of Georgia, Athens, GA, 1984.

4-1. *Soil Conservation in America.* American Farmland Trust, Washington, D.C., 1984.

4-2. D. Marsh. "When a Watershed is Gripped by Nonpoint Source Pollution." *All Riled Up: Nonpoint Source Pollution; The Point of No Return,* Wisconsin Dept. of Natural Resources, Madison, WI, no date.

4-3. D. Porter. "The West Eight Project." *Tennessee Conservationist.* LI(2):14-17, 1985.

5-1. O.S. Owen. *Natural Resources Conservation: An Ecological Approach,* 4th Ed., Macmillan Pub. Co., New York, NY, 1985.

5-2. J. Turk and A. Turk. *Environmental Science,* 3rd Ed., Saunders College Pub. Co., New York, NY, 1984.

5-3. S.N. Luoma. *Introduction to Environmental Issues.* Macmillan Pub. Co., New York, NY, 1984.

5-4. D.R. Lenat, L.A. Smock, and D.L. Penrose. "Use of Benthic Macroinvertebrates as Indicators of Environmental Quality." *Biological Monitoring for Environmental Effects,* by D.L. Wolf, Lexington Books, D.C. Heath and Co., Lexington, MA, 1980.

5-5. J. Carins, Jr. and K.L. Dickson. "A Simple Method for the Biological Assessment of the Effects of Waste Discharges on Aquatic Bottom Dwelling Organisms." *Jour. Water Pollution Control Fed.* 43(5):755-772, 1971.

6-1. J.B. Weber. "The Pesticide Scorecard." *Environ. Sci. & Technol.* 11(8):756-761, 1977.

6-2. B. Hileman. "Herbicides in Agriculture." *Environ. Sci. & Technol.* 16(12):645A-650A, 1982.

6-3. W.R. Mullison. "The Significance of Herbicides to Nontarget Organisms." Dow Chemical Co., Midland, MI, no date.

6-4. D. Pimentel and C.A. Edwards. "Pesticides and Ecosystems." *Bioscience,* 32(7):595-600, 1982.

6-5. A.W.A. Brown. *Ecology of Pesticides.* John Wiley & Sons, Inc., New York, NY, 1978.

6-6. F.L. Cross, Jr. *Handbook on Environmental Monitoring.* Technomic Pub. Co., Inc., Westport, CT, 1974.

6-7. J.R. Karr. "Assessment of Biotic Integrity Using Fish Communities." *Fisheries,* 6(6):21-27, 1981.

6-8. K.D. Fausch, J.R. Karr, and P.R. Yant. "Regional Application of an Index of Biotic Integrity Based on Stream Fish Communities." *Trans. Am. Fish Soc.*, 113:39-55, 1984.

6-9. R.J. Hall and D. Swineford. "Toxic Effects of Endrin and Toxaphene on the Southern Leopard Frog, *Rana sphenocephala.*" *Environ. Pollut.* (Series A) 23:53-65, 1980.

6-10. D.R. Bottrell. *Integrated Pest Management.* Council on Environmental Quality, Washington, D.C., 1979.

7-1. "Background Paper—Pollutants Causing Water Use Impairments: Animal Wastes." *Water Quality Training Facilitator's Guide.* U.S. Dept. of Agriculture, Soil Conservation Service, unpublished document, 1984.

7-2. *Agricultural Waste Management Field Manual.* U.S. Dept. of Agriculture, Soil Conservation Service, Engineering Div., Washington, D.C. 1975.

7-3. J.A. Moore, M.E. Grismer, S.R. Crane, and J.R. Miller. *Evaluating Dairy Waste Management Systems' Influence on Fecal Coliform Concentration in Runoff.* Agricultural Experiment Station Bulletin No. 658, Oregon State University, Corvallis, OR, 1982.

7-4. E. Moore, E. Janes, F. Kinsinger, K. Pitney, and J. Sainsbury. *Livestock Grazing Management and Water Quality Protection.* U.S. Environmental Protection Agency, Seattle, WA, 1979.

7-5. B.J. Nebel. *Environmental Science: The Way the World Works.* Prentice-Hall, Inc., Englewood Cliffs, NJ, 1981.

7-6. "Background Paper—Pollutants That Cause Water Use Impairments: Nutrients and Sediment." *Water Quality Training Facilitator's Guide.* U.S. Dept. of Agriculture, Soil Conservation Service, unpublished document, 1984.

7-7. *State-of-the-Art Review of Best Management Practices for Agricultural Nonpoint Source Control: Vol. I. Animal Waste.* North Carolina Agricultural Extension Service, Raleigh, NC, 1982.

7-8. J.A. Krivak. "Best Management Practices to Control Nonpoint Source Pollution from Agriculture." *Jour. Soil & Water Conserv.*, 33:161-166, 1978.

8-1. K. Kepler, D. Carlson and W.T. Pitts. *Pollution Control Manual for Irrigated Agriculture.* U.S. Environmental Protection Agency, Denver, CO, (EPA-908/3-78-002), 1978.

8-2. *Resources Conservation Act (RCA) Potential Problem Area II: Water Quality; Problem Statement and Objective Determination.* U.S. Dept. of Agriculture, Soil Conservation Service, Washington, D.C., 1979.

8-3. M.T. El-Ashry, J. van Shilfgaarde, and S. Schiffman. "Salinity Pollution from Irrigated Agriculture." *Jour. Soil & Water Conserv.*, 40(1):48-52, 1985.

8-4. W.Y. Bellinger and B.S. Bergendahl. *Highway Water Quality Monitoring Manual.* Federal Highway Administration, U.S. Dept. of Transportation, (Report No. FHWA-DP-43-2), Arlington, VA, 1979.

8-5. *"Best Management Practices" for Salinity Control in Grand Valley.* U.S. Environmental Protection Agency, (EPA-600/2-78-162), Ada, OK, 1978.

8-6. M.B. Holburt. "The Lower Colorado—A Salty River." *California Agriculture*, 38(10):6-8, 1984.

8-7. J. van Shilfgaarde. "Colorado River: Life Stream of the West." *Using Our Natural Resources*, Yearbook of Agriculture, U.S. Dept. of Agriculture, Washington, D.C., 1983.

8-8. T.W. Edminster and R.C. Reeve. "Drainage Problems and Methods." *Soil*, Yearbook of Agriculture, U.S. Dept. of Agriculture, Washington, D.C., 1957.

8-9. D.D. Chiras. *Environmental Science: A Framework for Decision Making*, 2nd Ed., Benjamin/Cummings Pub. Co., Inc., Menlo Park, CA, 1988.

8-10. *Saline-Seep Diagnosis, Control and Reclamation.* Agricultural Research Service, Conservation Research Report No. 30, U.S. Dept. of Agriculture, 1983.

A-1. W.M. Beck, Jr. "Suggested Method for Reporting Biotic Data." *Sewage and Industrial Wastes*, 27(10), 1955.

A-2. D. Wilson. *A Method for Determining Organic Enrichment of Surface Waters by Identification of Benthic Macroinvertebrates.* Tennessee Div. of Water Management, 1984.

B-1. C.M. Palmer. *Algae in Water Supplies: An Illustrated Manual on the Identification, Significance, and Control of Algae in Water Supplies.* U.S. Dept. of Health, Education and Welfare, Public Health Service, Div. of Water Supply and Pollution Control, Public Health Service Pub. No. 657, Washington, D.C., Reprinted 1962, no date.

B-2. P.M. Brady. *Pond Management for Sport Fishing in Arkansas.* U.S. Dept. of Agriculture, Soil Conservation Service, Little Rock, AR, 1981.

B-3. "Key to the Major Invertebrate Species of Streams." U.S. Dept. of Agriculture, Soil Conservation Service, no date.

B-4. D.S. Lee, C.R. Gilbert, C.H. Hocutt, R.E. Jenkins, D.E. McAllister, and J.R. Stauffer, Jr. *Atlas of North American Freshwater Fishes.* U.S. Fish and Wildlife Service and North Carolina State Museum of Natural History, North Carolina Biological Survey Pub. No. 1980-12, 1980.

B-5. *Standard Methods for the Examination of Water and Wastewater*, 16th Ed., Am. Public Health Assoc., Am. Water Works Assoc. and Water Pollution Control Federation, Washington, D.C., 1985.

B-6. A.B. Bottcher and L.B. Baldwin. "BMP Selector: General Guide for Selecting Agricultural Water Quality Practices." Institute of Food and Agricultural Sciences, University of Florida-Gainsville, Flordia Cooperative Extension Service Pub. N. SP-15, no date.

A Word of Thanks

Any effort to implement an idea involves a sound concept, a need to be filled, the dedication of many people, a little luck, and hard work. All these elements came together to complete the *Water Quality Indicators Guide: Surface Waters*. If any one of the above had been lacking, the guide would never have been finished. Each of the following people contributed in his or her own way. Four individuals deserve special thanks:

Dr. Patricia Bytnar Perfetti, Head of the Geoscience and Environmental Studies Department, and the Physics and Astronomy Department of the University of Tennessee at Chattanooga, researched and wrote much of the manuscript. Her assistance on the guide made it a reality. Selene Robinson of Trandes, Inc., not only typed the manuscript, but translated the field sheet format concept into a real, workable tool. Richard Francoeur, at the time in 1983 a graduating senior from Cornell University, made the first "cut" at a water quality indicator guide by doing initial research and developing the environmental cause/effect relationships that are present in the guide. Finally, my wife, Sandra, maintained her patience and good humor through many hours and days of proofreading.

Susan Alexander, EPA, Dallas
Malvern Allen, NTC*, PA
Mike Anderson, NE
Joseph Arruda, KS Dept. Health & Envrmt.
Donald Bivens, TN
Valerian Bohanty, NE
James Boykin, Ofc. Gov't & Pub. Aff., USDA
Bill Brown, CO
Gary Bullard, CA
John Burt, NTC, TX
Gerald Calhoun, MD
Sam Chapman, TX
Douglas Christensen, NHQ
Toby Clark, The Conservation Foundation, DC
Ellen Dietrich, PA
Thurston Dorsett, TN
Robert Drees, KS
Steven Dressing, EPA, NHQ
Paul DuMont, NHQ
Thomas Dumper, NTC, NE
Dennis Erinakes, NTC, TX
Robert Francis, NTC, PA
Robert Franzen, NTC, PA
Larry Goff, TN
Pat Graham, MO
Gary Gwinn, WV
Timothy Hall, FWS**, MD
Thomas Hamer, NE
Howard Hankin, TN
James Hannaham, DC Water Resources Resrch. Ctr.
Leaman Harris, EPA, Dallas
John Hassell, OK Conservation Commission
Steven Henningen, KS
Robert Higgins, KS
Patricia Hood-Greenberg, NHQ
Robert Hummel, KS
Thomas Iivari, NTC, PA
Barry Isaacs, PA
David Jones, MT
James Kaap, WI
Arnold King, NTC, TX
Susan King, EPA, Dallas
Judith Ladd, NHQ
Mary Landin, CE***, MS
Sarah Laurent, NHQ
Ronald Lauster, NM
Jerry Lee, TN
James Lewis, NHQ
Jeff Lozer, MD
Gary Margheim, NHQ
David McCalley, Univ. of No. Iowa
James Meek, EPA, NHQ
Daniel Merkel, NHQ
Milton Meyer, NHQ
Kent Milton, LA
David Moffitt, NTC, OR
Gerald Montgomery, TN
John Moore, NTC, PA
Robert Moorehouse, MA
Eldie Mustard, CO
Joel Myers, PA
James Newman, NHQ
Victor Payne, AL
Frank Resides, NTC, PA
Walter Rittall, NHQ
Larry Robinson, SC
Marc Safley, NHQ
Donald Schuster, MN
Jane Sisk, Calloway Co. Pub. Schools, KY
Daniel Smith, NHQ
Donald Snethen, KS Dept. Health & Envmt.
Frank Sprague, NTC, TX
Lyle Steffan, CA
James Stiebing, EPA, Dallas
Billy Teels, NHQ
Mark Waggoner, MN
Clive Walker, NHQ
Gerald Welsh, NHQ
Robert Wengrzynek, ME

Charles R. Terrell
National Water Quality Specialist
Soil Conservation Service, Washington, D.C. 20013

* SCS National Technical Center
** U.S. Fish & Wildlife Service
*** U.S. Army Corps of Engineers

Appendix E

Conservation and Best Management Practices

List of conservation and best management practices (BMP's) that can be employed to reduce or eliminate nonpoint source water pollution problems.

1. **Access Road**—A road located and constructed to provide needed access, but built with soil conservation measures to prevent soil erosion caused by vehicular traffic or animal travel.
2. **Alternative Pesticides**—Pesticides other than chemical types traditionally used on a crop.
3. **Bedding**—Plowing, blading, or otherwise elevating the surface of flat land into a series of broad, low ridges separated by shallow, parallel channels.
4. **Biological Control Methods**—Use of organisms or biological materials to control crop pests. Integrated Pest Management (IPM) is an example of biological control that can reduce the amounts of chemical pesticides needed to grow a crop.
5. **Brush Management**—Management and manipulation of brush to improve or restore plant cover quality in reducing soil erosion.
6. **Chiseling and Subsoiling**—Loosening the soil to shatter compacted and restrictive layers to improve water quality, infiltration and root penetration, and reduce surface water runoff.
7. **Conservation Cropping**—Growing crops in combination with needed cultural and management measures to improve the soil and protect it during erosion periods. Practices include cover cropping and crop rotation, and providing vegetative cover between crop seasons.
8. **Conservation Cropping Sequence**—A sequence of crops designed to provide adequate organic residue to maintain and improve soil tilth.
9. **Conservation Tillage**—In producing a crop, limiting the number of cultural operations to reduce soil erosion, soil compaction, and energy use. Usually involves an increase in the use of herbicides.
10. **Contour Farming**—Farming sloped land on the contour to reduce erosion, control water flow, and increase infiltration.
11. **Contour Orchard and Other Fruit Areas**—Planting orchards, vineyards, or small fruits, so all cultural operations are done on the contour.
12. **Correct Fertilizer Container Disposal**—Following accepted methods for fertilizer container disposal, keeping containers out of sinkholes, creeks, and other places adjacent to water to reduce the amount of fertilizer that reaches waterways.
13. **Correct Pesticide Container Disposal**—Following accepted methods for pesticide container disposal, keeping containers out of sinkholes, creeks, and other places adjacent to water to reduce the amount of pesticide that reaches waterways.
14. **Cover and Green Manure Crops**—Use of close-growing grasses, legumes, or small grain for seasonal soil protection and improvement.
15. **Critical Area Planting**—Planting vegetation to stabilize the soil and reduce erosion and runoff.
16. **Crop Residue Use**—Leaving plant residues after harvest to protect cultivated fields during critical erosion periods when the ground would otherwise be bare.
17. **Crop Rotation**—Planting different crops in successive seasons in the same field. Procedure can reduce pesticide loss significantly. There are some indirect costs if less profitable crops are alternated.
18. **Debris Basin**—A barrier or berm constructed across a watercourse or at other suitable locations to act as a silt or sediment catchment basin.
19. **Deferred Grazing**—Postponing grazing for a prescribed period to improve vegetative conditions and reduce soil loss.
20. **Diversion**—Channels constructed across a slope to divert runoff water and help control soil erosion, and having a mound or ridge along the lower side of the slope.
21. **Drainage Land Grading**—Reshaping the surface of land to improve surface drainage and/or water distribution.
22. **Emergency Tillage**—Roughening soil surfaces by methods, such as listing, ridging, duck-footing, or chiseling. Procedure is done as a temporary protection measure.
23. **Farmstead and Feedlot Windbreak**—A strip or belt of trees or shrubs, established next to a farmstead or feedlot to reduce wind speed and protect soil resources.
24. **Fencing**—Enclosing an environmentally sensitive area of land or water with fencing to control access of animals or people.
25. **Field Border**—A border or strip of permanent vegetation, established at field edges to control soil erosion and slow, reduce, or eliminate pollutants from entering an adjacent watercourse or water body.
26. **Field Windbreak**—A strip or belt of trees or shrubs, established in or adjacent to a field, to reduce wind speed and protect soil resources.
27. **Filter Strip**—A strip or section of land in permanent vegetation, established downslope of agricultural operations to control erosion and slow, reduce, or eliminate pollutants from entering an adjacent watercourse.
28. **Fishpond Management**—Developing or improving impounded water to produce fish for consumption or recreation.
29. **Grade Stabilization Structure**—A structure to stabilize a streambed or to control erosion in natural or constructed channels.
30. **Grasses and Legumes in Rotation**—A conservation cropping system that establishes and maintains grasses and/or legumes for a definite number of years.
31. **Grazing Land Mechanical Treatment**—Renovating, contouring, furrowing, pitting, or chiseling native grazing land by mechanical means to improve plant cover and water availability.
32. **Heavy-Use Area Protection**—Establishing vegetative cover or installing structures to stabilize heavily used areas.

33. **Hillside Ditch**—A channel constructed to control the water flow and erosion by diverting runoff to a protected outlet.

34. **Integrated Pest Management Program**—Use of organisms or biological materials for effective pest control with reduction in amounts of pesticides used. "Scouting" of insect pest populations is necessary to determine when pest management actions are necessary to reduce pests.

35. **Irrigation Field Ditch**—A permanently lined irrigation ditch that conveys water from a supply source to fields, preventing erosion, infiltration, or degradation of water quality.

36. **Irrigation Water Conveyance**—A pipeline or lined waterway constructed to prevent erosion and loss of water.

37. **Irrigation Water Management**—Determining and controlling the rate, amount, and timing of irrigation water applied to crops to minimize soil erosion, runoff, and fertilizer and pesticide movement.

38. **Land Absorption Areas and Use of Natural or Constructed Wetland Systems**—Providing adequate land absorption or wetland areas downstream from agricultural areas so that soil and plants receive and treat agricultural nonpoint source pollutants.

39. **Listing**—Plowing and planting done in the same operation. Plowed soil is pushed into ridges between rows, and seeds are planted in the furrows between the ridges.

40. **Livestock Exclusion**—Excluding livestock from environmentally sensitive areas to protect areas from induced damages. Also, excluding livestock from areas not intended for grazing.

41. **Precision Application Rates**—Within a particular field, applying precise amounts of fertilizer and pesticide according to the soil/plant needs in specific parts of the field. Generally, lower rates can be applied, especially where tests show residues are present from previous applications.

42. **Managing Aerial Pesticide Applications**—Having pesticides applied when winds are low and when they are in a direction away from watercourses and riparian areas. This can reduce contamination in these nontarget areas.

43. **Mechanical Weed Control Methods**—Using mechanical or biological, instead of chemical, weed control can reduce substantially the need for chemicals. Costs will have to be carefully computed to make the operation economically feasible.

44. **Minimizing Number of Irrigations**—Carefully monitoring crop water needs and soil water availability minimizes the number of irrigations necessary to produce a crop. This may yield higher profits at harvest and reduce water pollution and soil erosion.

45. **Mulching**—Applying plant residues or other suitable materials to the soil surface reduces evaporation, water runoff, and soil erosion. Plastic sheeting can increase runoff, but will reduce nutrient leaching.

46. **No-till or Zero-tillage**—Tilling the soil with minimal disturbance and utilizing a fluted colter or double-disk opener ahead of the planter shoe to cut through untilled residues of the previous crop.

47. **Optimizing Crop Planting Time**—Planting a crop at a time other than when the crop's specific pest enemies would be present can reduce the need for pesticides and lower costs.

48. **Optimizing Date of Application**—Changing a pesticide application date to avoid impending rain or winds can improve effectiveness of the pesticide application and avoid environmental problems. Application can only be done when pest control effectiveness is not adversely affected. Process involves little or no cost.

49. **Optimizing Pesticide Formulations**—Pesticides come in several formulations with different half-lives. If a formulation with a shorter half-life than one normally used by the farmer is chosen, the pesticide will be less available to cause environmental damage. Also, some formulations require fewer applications for the same pest protection, so costs are reduced and less is available to the environment.

50. **Optimizing Pesticide Placement**—Direct application of a pesticide on the field and plants rather than aerial spraying is more effective, reduces costs, and protects nearby environments from accidental spraying.

51. **Optimizing Time of Day For Application**—Applying pesticide at times of low winds, often early and late in the day, can reduce amounts needed for the crop, reduce costs, and reduce pesticide that could adversely affect adjacent environments.

52. **Pasture and Hayland Management**—Proper treatment, including fertilizing, aerating, and harvesting can protect soil and reduce water loss.

53. **Phreatophyte Water Losses**—Elimination of nonbeneficial uses of water by phreatophytes (plants getting water from deep roots) not only lessens the concentration of salts through transpiration, but conserves water as well. Lowering the water table and developing mechanical and chemical techniques for elimination of phreatophytes ensures more efficient water use and minimizes salt hazards.

54. **Planned Grazing Systems**—A system in which two or more grazing units are alternately grazed and rested from grazing in a planned sequence to improve forage production, maintain vegetative cover, retain animal wastes on the land, and protect animals from polluted waters.

55. **Plant Between Rows in Minimum Tillage**—Applicable only to row crops in non-plow-based tillage; may reduce amounts of pesticides necessary.

56. **Plow-Plant**—Crop is planted directly into plowed ground with secondary tillage. This system increases infiltration and water storage.

57. **Pond**—A water impoundment made by constructing a dam or embankment or by excavating a pit or "dugout."

58. **Pond Sealing or Lining**—Installing a fixed lining of impervious material or treating the soil in a pond to reduce or prevent excessive water loss.

59. **Precision Land Forming**—Reshaping the surface of land to planned grades to give effective and efficient water movement.

60. **Proper Fertilizer Applications**—Selecting the proper time and method of fertilizer application to reduce losses through leaching and soil erosion, and ensure adequate crop nutrition.

61. **Proper Grazing Use**—Having no more animal units than will allow grazing areas to maintain sufficiently healthy, productive vegetative cover to protect the soil from eroding and protect the water quality of adjacent watercourses.

62. **Proper Timing of Irrigation Sprinklers**—Using irrigation equipment when plants need moisture, and controlling the amount of moisture delivered to the plants by avoiding over-irrigating to conserve water, protect soil from eroding, and protect the water quality of adjacent watercourses.

63. **Pumped Well Drain**—A well sunk into an aquifer to pump water to lower the prevailing water table.

64. **Pumping Plant for Water Control**—A pumping facility installed to transfer water for a conservation need.

65. **Range Seeding**—Establishing adapted plants on rangeland to reduce soil and water loss and produce more forage.

66. **Reducing Excessive Insecticide Treatment**—Applying exactly the correct amounts of insecticide recommended by the manufacturer for the crop and soil types. Refined predictive techniques required, such as computer forecasting.

67. **Reduction of Weed Growth**—Reducing number of weed plants to reduce water loss from evapotranspiration.

68. **Reduction or Elimination of Irrigation of Marginal Lands**—Taking irrigated marginally productive lands out of production to reduce water losses and salt pollution.

69. **Regulated Runoff Impoundment**—Retention or detention of water with infiltration prior to discharge to reduce runoff quantity, retain nutrients and pesticides, and prevent pollutants from reaching watercourses.

70. **Regulating Water in Drainage Systems**—The use of water-control structures to control the removal of surface runoff waters or subsurface flows.

71. **Reservoir Evaporation**—Controlling, through design or practices, the evaporation rate of water from reservoirs. If not controlled, evaporation tends to increase the salt content of the reservoir waters.

72. **Resistant Crop Varieties**—Use of plant varieties that are resistant to insects, nematodes, diseases, salt, etc.

73. **Return Flow Regulation**—Regulating the type and quantity of water return flows as a means of maintaining and improving irrigation water quality.

74. **Ridge Tillage**—Tillage producing a row configuration similar to listing, but planting is done on the ridges year after year with no seedbed preparation preceding planting.

75. **Rock Barrier**—A rock retaining wall, constructed across the slope, forming and supporting a bench terrace to control the flow of water on sloping land.

76. **Roof Runoff Management**—A facility for collecting, controlling, and disposing of rainfall/snowmelt runoff water from roofs. It keeps animal holding areas free of excess water and helps to maintain water quality of adjacent watercourses.

77. **Row Arrangement**—Establishing crop rows on planned grades and lengths to provide drainage and erosion control.

78. **Runoff Management System**—A system for controlling excess runoff from a development site during and after construction operations.

79. **Sediment Basin**—A basin constructed to collect and store sediment from runoff waters associated with nonpoint source pollutants.

80. **Slow Release Fertilizer**—Applying fertilizers that release nitrogen slowly to soil and plants, to minimize rapid nitrogen losses from soils prone to leaching.

81. **Soil Testing and Plant Analysis**—Testing soils and determining plant fertilizer requirements to avoid overfertilization and subsequent nutrient losses to runoff water.

82. **Split Applications of Nitrogen**—"Splitting" or dividing a set amount of fertilizer into two or more applications in the same season for the same crop.

83. **Spring Development**—Improving springs and water seeps by excavating, cleaning, capping, or providing collection and storage facilities for the water.

84. **Spring Nitrogen Fertilizer Application**—Applying nitrogen fertilizer in the spring, instead of autumn, to avoid fertilizer losses from heavy late winter and early spring runoff events.

85. **Streambank Protection**—By vegetative or structural means, stabilizing and protecting banks of watercourses, lakes, estuaries, or excavated channels against scour and erosion.

86. **Strip Tillage**—A narrow strip, tilled with a rototiller gang or other implement. Seed is planted in the same operation.

87. **Stripcropping**—Growing crops in a systematic arrangement of strips or bands to reduce water and wind erosion.

88. **Stripcropping, Contour**—Growing crops on the contour to reduce erosion and control water.

89. **Stripcropping, Field**—Planting large sections or entire fields in a systematic arrangement to help control erosion and runoff on sloping cropland where contour stripcropping is not a practical method.

90. **Structure for Water Control**—A structure to control the water stage, discharge, distribution, delivery, or direction of water flow in open channels or water use areas.

91. **Subsurface Drain**—A conduit, such as tile or plastic pipe, installed beneath the ground surface to control water levels for increased production. Net runoff and leaching are reduced, but nitrate concentrations may be increased.

92. **Surface Drainage**—A conduit, such as tile, pipe, or tubing, installed beneath the ground surface to collect and/or convey drainage water.

93. **Surface Roughening**—Roughening the soil surface by ridge or clod-forming tillage.

94. **Sweep Tillage**—Using a "sweep" on small-grain stubble to kill early fall weeds. The practice shatters and lifts the soil, thus enhancing infiltration while leaving residue in place.

95. **Terrace**—An earth embankment, channel, or a combination ridge and channel constructed across a slope to control runoff.

96. **Timing and Placement of Fertilizers**—Delaying timing or using proper placement of fertilizers for maximum utilization by plants and minimum fertilizer leaching or movement by surface runoff.

97. **Tree Planting**—To establish or reinforce a stand of trees to conserve soil and moisture and help protect water leaving agricultural areas by "filtering" pollutants from the water flow.

98. **Trickle Irrigation**—Using trickle irrigation equipment to deliver small quantities of water to irrigate crops.

99. **Trough or Tank**—Locating watering facilities a reasonable distance from watercourses and dispersing the facilities to encourage uniform grazing and to reduce livestock concentrations, particularly near watercourses.

100. **Underground Outlet**—A water outlet, placed underground to dispose of excess water without causing damage by erosion or flooding.

101. **Uniformity of Irrigation Water Quality**—Uniform irrigation water quality can be achieved through water flow regulation by controlling the release of water from storage reservoirs.

102. **Waste Management System**—A planned system to manage animal wastes in a manner that does not degrade air, soil, or water resources. Often wastes are collected in storage or treatment impoundments, such as ponds, lagoons, or stacking facilities.

103. **Waste Storage Pond**—An impoundment for temporary storage of animal or other agricultural waste.

104. **Waste Storage Structure**—A fabricated structure for the temporary storage of animal wastes or other organic agricultural wastes.

105. **Waste Treatment Lagoon**—An impoundment for biological treatment of animal or other agricultural waste.

106. **Waste Utilization**—Using wastes for fertilizer or other purposes in a manner which improves the soil and protects water resources. May also include recycling of waste solids for animal feed supplement.

107. **Water and Sediment Control Basin**—An earth embankment or a combination ridge and channel to form a sediment trap and a water detention basin to prevent soil erosion losses and improve water quality.

108. **Water Supply Dispersal**—A well which is constructed or improved to provide water for irrigation and livestock and which enhances natural livestock distribution or improved vegetative cover.

109. **Water Spreading**—Diverting or collecting runoff and spreading it over relatively flat areas.

APPENDIX F

Soil Conservation Service

Water Quality Indicators Guide: Surface Waters

Field Sheets

Note: Copy the assessment and field sheets before proceeding! Write on the copies.

Part 1: Background information for the watershed assessment needs to be completed only once for each watershed. The assessment gives general information about the watershed and may serve for several watercourses or water bodies within the watershed. The on-farm (ranch) water assessment will have to be completed for each farm or ranch evaluated.

Part 2: Field sheet selection of nonpoint source pollutants will have to be completed for each watercourse or water body evaluated. This preliminary decision about pollutants will determine which field sheets will need to be completed (sediment, animal wastes, nutrients, pesticides, or salts).

Part 1: Background Information

Watershed Assessment

Evaluator's Name _____ Date_____

Location _____

State/County _____ Township/Range _____

Watershed (basin) _____ Subwatershed _____
(if applicable)

_____

Watercourses/Water Bodies

1. Size of watershed_____

2. Number of major watercourses_____

3. Watercourse names_____

4. Types of watercourses_____ ephemeral*_____ intermittent* or _____ perennial*

5. Average watercourse gradient (feet per mile) _____

6. Watercourse bottom (predominant type):
   ☐ bedrock ☐ boulder ☐ cobble ☐ gravel ☐ sand ☐ silt-clay ☐ organic

7. Frequency of flooding: ☐ none ☐ rare ☐ occasional ☐ frequent

8. Watercourse channel alteration: ☐ dredged ☐ channelized ☐ other (alteration date if known)

9. Watercourse primary uses:
   (1) ☐ Domestic drinking water supply
   (2) ☐ Industrial water supply
   (3) ☐ Agricultural water supply: ☐ Irrigation ☐ Livestock ☐ Other (explain)_____

_____

   (4) ☐ Recreation: ☐ Swimming ☐ Fishing ☐ Other (explain)_____

   (5) Other uses (explain) _____

_____
*Stream definitions: See Glossary in Appendix C.

Watershed Assessment (Continued)

10. Water use impairments (watercourses): Are there water use impairments or restrictions of the watercourses in the watershed? Is there something "wrong" with the water? Has it been degraded by acid mine drainage, industrial discharge, etc.?

    ☐ No. ☐ Yes. If Yes, explain. The impairment(s) is/are due to:

    ☐ Agricultural runoff   ☐ Logging runoff   ☐ Inadequate or overloaded wastewater treatment facilities
    ☐ Industrial discharge   ☐ Urban or other construction ☐ Irrigation problems
    ☐ Mining runoff   ☐ Failing septic tanks   ☐ Other (explain) _____

    If the impairment(s) is/are due to a combination of factors, what is your best estimate of the relative contribution of agricultural operations to the watercourse impairment(s)?

    ☐ Total   ☐ About half
    ☐ Most   ☐ Small portion

11. Names of major ponds or lakes _____

12. Names of minor ponds or lakes _____

13. Sizes of individual ponds or lakes in acres _____

14. Primary uses of pond or lake water:

    (1) ☐ Domestic drinking water supply.

    (2) ☐ Industrial water supply.

    (3) ☐ Agricultural water supply: ☐ Irrigation ☐ Livestock ☐ Other (explain)_____

    (4) ☐ Recreation: ☐ Swimming ☐ Fishing ☐ Other (explain)_____

    (5) ☐ Other uses (explain) _____

15. Water use impairment (ponds or lakes): Are there any water use impairments or restrictions of the ponds or lakes in the watershed? Is there something "wrong" with the water? Has it been degraded by acid mine drainage, industrial discharge, etc?

    ☐ No. ☐ Yes. If Yes, explain. The impairment(s) is/are due to:

    ☐ Agricultural runoff   ☐ Logging runoff   ☐ Inadequate or overloaded wastewater treatment facilities
    ☐ Industrial discharge   ☐ Urban or other construction ☐ Irrigation problems
    ☐ Mining runoff   ☐ Failing septic tanks   ☐ Other (explain) _____

If the impairment(s) is/are due to a combination of factors, what is your best estimate of the relative contribution of agricultural operations to the pond/lake impairment(s)?

☐ Total ☐ About half

☐ Most ☐ Small portion

16. Land uses in watershed: (check appropriate categories)

    1. Farming: ☐ Pasture/grazing ☐ Dryland cropping ☐ Irrigated cropping ☐ Woods

        Other (explain) _____

    2. Urban areas: ☐ Homes ☐ Stores ☐ Other (explain)_____

    3. Industrial areas: ☐ Factories ☐ Small shops ☐ Other (explain)_____

    4. Mining: ☐ Surface ☐ Deep ☐ Other (explain)_____

    5. Logging: ☐ Clearcut ☐ Selective cut ☐ Other (explain)_____

    6. Other uses (explain)_____ (e.g., sanitary landfill)

On-Farm (Ranch) Water Assessment

Surface Watercourses

1. Number of watercourses (streams or drainage ditches) _____

2. Types of watercourses: ☐ Perennial ☐ Intermittent ☐ Ephemeral. (If applicable, check more than one).

3. Location of watercourses on property* _____

Wetlands

1. Acres of wetlands _____

2. Location of wetlands on property* _____

3. Uses: ☐ Sediment sink ☐ Water storage ☐ Flood control ☐ Irrigation

   ☐ Other (explain) _____

Ponds

1. Farm ponds: ☐ Rare ☐ Common ☐ Abundant (Number_____)

2. Other water bodies: Number _____; Surface area (natural lakes and impounded bodies) _____

3. Uses _____

4. Are any of the uses impaired? ☐ No. ☐ Yes. If yes, what type of impairment? _____

5. Location on property* _____

Ground Water

1. Number of springs (wet weather or year-round): ☐ None ☐ Rare ☐ Common ☐ Abundant

2. Total number of wells _____

3. Population served _____

4. Primary uses of ground water:

   ☐ Domestic water supply   ☐ Industrial water supply   ☐ Recreation

   ☐ Fish and aquatic life   ☐ Irrigation

   ☐ Livestock watering & wildlife   ☐ Other (explain)_____

5. Number of sinkholes on or near property: ☐ None ☐ Rare ☐ Common ☐ Abundant

6. Location of ground water features on property* _____

---

*Optional: If feasible and not already present, add these locations to the farm's map.

Part 2: Field Sheet Selection of Nonpoint Source Pollutants

Watercourses

Evaluator's Name _____ Date_____

Location _____

State/County_____ Township/Range _____

Watershed (basin) _____ Subwatershed _____
                                                                                            (if applicable)

Watercourse location _____

Note: If there is a natural or constructed watercourse on or near the farmer or rancher's property, complete this form for a preliminary decision about nonpoint source pollutants. If your answer is "Can't Tell" or "Yes," you must complete the field sheets for that particular pollutant.

Probable Cause

1. Is the watercourse bottom coated with sediment? *Or* is there evidence that the watercourse bed is aggrading or degrading? *Or* has flooding increased over the last several years? *Or* is there evidence of bank sloughing? *Or* is the water turbid or muddy after a storm event?

   ☐ CAN'T TELL   ☐ YES   ☐ NO   (Field Sheets 1A, 1B)   **Sediment**

2. Do you see or smell evidence of manure in or along the watercourse? *Or* is there evidence of bank trampling? *Or* is the watercourse bottom coated black or with a whitish or grayish "cottony" mold?

   ☐ CAN'T TELL   ☐ YES   ☐ NO   (Field Sheets 2A, $2B_1$, $2B_2$)   **Animal Wastes**

3. At low flow is the color of the water greenish? *Or* is there an increase in rooted aquatic vegetation or seasonal algal blooms that can be linked to the timing of fertilizer application?

   ☐ CAN'T TELL   ☐ YES   ☐ NO   (Field Sheets 3A, 3B)   **Nutrients**

4. Is there evidence of leaf-burn or a sudden dieback of vegetation that does not seem to be due to natural causes? Has this happened after pesticide application? *Or* has fish productivity declined or a fishery been degraded from cold or warm water sport fish to predominately rough (trash) fish? *Or* has a fish kill occurred in an apparently fertile watershed? *Or* do fish avoid this particular reach of the watercourse or exhibit strange or erratic behavior such as gulping for air, swimming in circles, jumping out of water, etc?

   ☐ CAN'T TELL   ☐ YES   ☐ NO   (Field Sheets 4A, 4B)   **Pesticides**

5. Is the watercourse reach located in a naturally occurring salt-laden geologic area and far downstream from the headwaters? *Or* is there evidence of white salt crust on the watercourse banks?

   ☐ CAN'T TELL   ☐ YES   ☐ NO   (Field Sheets 5A, $5B_1$, $5B_2$)   **Salts**

After completing this preliminary decision about pollutants, complete the appropriate field sheets.

Water Bodies (Ponds or Lakes)

If there is a natural or constructed pond or lake on or near the farmer or rancher's property, complete this form for a preliminary decision about nonpoint source pollutants. If your answer is "Can't tell" or "Yes," you must complete the field sheets for that particular pollutant.

Water body location **Probable Cause**

1. Is there evidence of sediment from field erosion or bank erosion getting into the pond from rills or gullies; muddiness after a large storm? *Or* has the pond changed in size over the years? (Pond surface area may become smaller or larger as sedimentation occurs.)

    ☐ CAN'T TELL ☐ YES ☐ NO (Field Sheets 1A, 1B) **Sediment**

2. Is there visual or olfactory evidence of manure in or around the pond? *Or* is the pond covered with an organic ooze or black mayonnaise-like coating?

    ☐ CAN'T TELL ☐ YES ☐ NO (Field Sheets 2A, $2B_1$, $2B_2$) **Animal Waste**

3. Have there been seasonal algal blooms or fish kills or evidence of oxygen depletion (fish gulping for air near the surface of the water at dawn)? *Or* is there evidence of greener, more robust vegetation along the pond edge? *Or* is the pond choked with vegetation?

    ☐ CAN'T TELL ☐ YES ☐ NO (Field Sheets 3A, 3B) **Nutrients**

4. Is there evidence of leaf-burn or a sudden dieback of vegetation that does not seem to be due to natural causes? Has this occurred after pesticide application? *Or* has fish productivity declined or a fish kill occurred in an apparently fertile watershed with good pond management practices? *Or* has abnormal fish behavior been observed, such as uncoordinated movements, convulsive darting movements, erratic swimming up and down or in a small circle, sluggish movements alternating with jumping out of the water, or difficulty in respiration?

    ☐ CAN'T TELL ☐ YES ☐ NO (Field Sheets 4A, 4B) **Pesticides**

5. Is the pond or lake located in a naturally occurring salt-laden geologic area and far downstream from the headwaters? *Or* has there been a significant amount of irrigation drainage to the pond and a need to increase the number of evaporation ponds in a given area? *Or* are pond shorelines covered with white, crusty salt deposits?

    ☐ CAN'T TELL ☐ YES ☐ NO (Field Sheets 5A, $5B_1$, $5B_2$) **Salts**

After completing this preliminary decision about pollutants, complete the appropriate field sheets.

Sediment Page 1 of 2

FIELD SHEET 1A: SEDIMENT
INDICATORS FOR RECEIVING WATERCOURSES AND WATER BODIES

Evaluator _____ Water Body Location _____ County/State _____ Date _____

Water Body Evaluated _____ Total Score/Rank _____

(Circle one number among the four choices in each row which BEST describes the conditions of the watercourse or water body being evaluated. If a condition has characteristics of two categories, you can "split" a score.)

Rating Item	Excellent	Good	Fair	Poor
1. Turbidity (best observed immediately following a storm event)	-- Clear or very slightly muddy after storm event. -- Objects visible at depths greater than 3 to 6 ft. (depending on water color). -- OTHER **9**	-- What is expected for properly managed agricultural land in your region. -- A little muddy after storm event but clears rapidly. -- Objects visible at depths between 1½ to 3 ft. (depending on water color). -- OTHER **7**	-- A considerable increase in turbidity for your region. -- Considerable muddiness after a storm event. Stays slightly muddy most of the time. -- Objects visible to depths of ½ to 1½ ft. (depending on water color). -- OTHER **3**	-- A significant increase in turbidity for your region. -- Very muddy—sediment stays suspended most of the time. -- Objects visible to depths less than ½ ft. (depending on water color). -- OTHER **0**
2. Bank stability in your viewing area	-- Bank stabilized. -- No bank sloughing. -- Bank armored with vegetation, roots, brush, grass, etc. -- No exposed tree roots. -- OTHER **10**	-- Some bank instability. -- Occasional sloughing. -- Bank well-vegetated. -- Some exposed tree roots. -- OTHER **7**	-- Bank instability common. -- Sloughing common. -- Bank sparsely vegetated. -- Many exposed tree roots & some fallen trees or missing fence corners, etc. -- Channel cross-section becomes more U-shaped as opposed to V-shaped. -- OTHER **4**	-- Significant bank instability. -- Massive sloughing. -- No vegetation on bank. -- Many fallen trees, eroded culverts, downed fences, etc. -- Channel cross-section is U-shaped and stream course or gully may be meandering. -- OTHER **1**
3. Deposition (Circle a number in only A, B, C, or D)	SELECT 3A OR 3B OR 3C OR 3D			
3A. Rock or gravel streams OR	A. For rock and gravel bottom streams: -- Less than 10% burial of gravels, cobbles, and rocks. -- Pools essentially sediment free. **9**	A. For rock and gravel bottom streams: -- Between 10% & 25% burial of gravels, cobbles, & rocks. -- Pools with light dusting of sediment. **7**	A. For rock & gravel bottom streams: -- Between 25% and 50% burial of gravels, cobbles and rock. -- Pools with a heavy coating of sediment. **3**	A. For rock & gravel bottom streams: -- Greater than 50% burial of gravels, cobbles and rocks. -- Few if any deep pools present. **1**

3B. Sandy bottom streams	B. For sandy streambeds: -- Sand bars stable and completely vegetated. -- No mudcaps or "drapes" (coverings of fine mud). -- No mud plastering of banks; exposed parent material. -- No deltas. -- OTHER 9	B. For sandy streambeds: -- Sand bars essentially stable and well, but not completely, vegetated. -- Occasional mudcaps or "drapes." -- Some mud plastering of banks. -- Beginnings of delta formation. 7	B. For sandy streambeds: -- Sand bars unstable with sparse vegetation. -- Mudcaps or "drapes" common. -- Considerable mud plastering of banks. -- Significant delta formation. 3	B. For sandy streambeds: -- Sand bars unstable and actively moving with no vegetation. -- Extensive mudcaps or "drapes." -- Extensive mud plastering of banks. -- Extensive deltas. 1
3C. Mud-bottom streams OR	C. For mud bottom streams: -- Dark brown/black tanic-colored water (due to presence of lignins and tanins). -- Abundant emergent rooted aquatics or floating vegetation. -- OTHER 9	C. For mud bottom streams: -- Dark brown colored water. -- OTHER 7	C. For mud bottom streams: -- Medium brown water, muddy bottom. -- OTHER 3	C. For mud bottom streams: -- Light brown colored, very muddy bottom. -- OTHER 1
3D. Ponds	-- Ponds essentially sediment free. -- No reduction in pond storage capacity. -- OTHER 9	-- Ponds with light dusting of sediment. -- Very little loss in pond storage capacity. -- OTHER 7	-- Ponds with a heavy coating of sediment. -- Some measurable loss in pond storage capacity. -- OTHER 3	-- Ponds filled with sediment. -- Significant reduction in pool storage capacity. -- OTHER 1

Sediment Page 2 of 2

FIELD SHEET 1A: SEDIMENT, Continued
INDICATORS FOR RECEIVING WATERCOURSES AND WATER BODIES

Rating Item	Excellent	Good	Fair	Poor
4. Type and amount of aquatic vegetation & condition of periphyton (plants, growing on other plants, twigs, stones, etc.)	-- Periphyton bright green to black. Robust. -- Abundant emergent rooted aquatics or shoreline vegetation. -- In ponds, emergent rooted aquatics (e.g. cattails, arrowhead, pickerelweed, etc.) present, but in localized patches. -- OTHER 9	-- Periphyton pale green and spindly. -- Emergent rooted aquatics or shoreline vegetation common. -- In ponds, emergent rooted aquatics common, but confined to well-defined band along shore. -- OTHER 7	-- Periphyton very light colored or brownish and significantly dwarfed. Sparse vegetation. -- In ponds, emergent rooted aquatics abundant in wide bank; encroachment of dry land species (grasses, etc.) along shore. -- OTHER 5	-- No periphyton. -- No vegetation. -- In ponds, emergent rooted aquatics predominant with heavy encroachment of dry land species. -- OTHER 2
OPTIONAL: 5. Bottom stability of streams	-- Stable. -- Less than 5% of stream reach has evidence of scouring or silting. -- OTHER 9	-- Slight fluctuation of streambed up or down (aggradation or degradation). -- Between 5-30% of stream reach has evidence of scouring or silting. -- OTHER 7	-- Considerable fluctuation of streambed up or down (aggradation or degradation). -- Scoured or silted areas covering 30-50% of evaluated stream reach. -- Flooding more common than usual. -- More stream braiding than usual for region. -- OTHER 3	-- Significant fluctuation of streambed up or down (aggradation or degradation). -- More than 50% of stream reach affected by scouring or deposition. -- Flooding very common. -- Significantly more stream braiding than usual for region. -- OTHER 1

OPTIONAL:

6. Bottom dwelling aquatic organisms	:-- Intolerant species occur: mayflies, stoneflies, caddisflies, water penny, riffle beetle and a mix of tolerants. :-- High diversity. :-- OTHER	:-- A mix of tolerants: shrimp, damselflies, dragonflies, black flies. :-- Intolerants rare. :-- Moderate diversity. :-- OTHER	:-- Many tolerants (snails, shrimp, damselflies, dragon flies, black flies). :-- Mainly tolerants and some very tolerants. :-- Intolerants rare. :-- Reduced diversity with occasional upsurges of tolerants, e.g. tube worms and chironomids. :-- OTHER	:-- Only tolerants or very tolerants: midges, craneflies, horseflies, rat-tailed maggots, or none at all. :-- Very reduced diversity; upsurges of very tolerants common. :-- OTHER
	9	7	3	1

TOTAL [   ]

1. Add the circled Rating Item scores to get a total for the field sheet.
2. Check the ranking for this site based on the total field score. (Check "excellent" if the score totals at least 32. Check "good" if the score falls between 21 and 31, etc.). Record your total score and rank (excellent, good, etc.) in the upper right-hand corner of the field sheet. If a Rating Item is "fair" or "poor," complete Field Sheet 1B.

RANKING				
	Excellent (32-37) [ ]	Good (21-31) [ ]	Fair (9-20) [ ]	Poor (8 or less) [ ]
OPTIONAL RANKING (with #5 OR #6)	Excellent (40-46) [ ]	Good (26-39) [ ]	Fair (11-25) [ ]	Poor (10 or less) [ ]
OPTIONAL RANKING (with #5 AND #6)	Excellent (48-55) [ ]	Good (31-47) [ ]	Fair (13-30) [ ]	Poor (12 or less) [ ]

Sediment

FIELD SHEET 1B: SEDIMENT
INDICATORS FOR CROPLAND, HAYLAND OR PASTURE

Evaluator _____ County/State _____ Date _____

Field Evaluated _____ Field Location _____ Total Score/Rank _____

(Circle one number among the four choices in each row which BEST describes the conditions of the field or area being evaluated. If a condition has characteristics of two categories, you can "split" a score.)

Rating Item	Excellent	Good	Fair	Poor	Practices from Appendix E
1. Erosion Potential	-- Not significant. -- Less than T (tolerance); little sheet, rill, or furrow erosion. -- No gullies. -- OTHER 10	-- Some erosion evident. -- About T; some sheet, rill, or furrow erosion. -- Very few gullies. -- OTHER 7	-- Moderate erosion. -- T to 2T. -- Gullies or furrows from heavy storm events obvious. -- OTHER 3	-- Heavy erosion. -- More than 2T. -- Many gullies or furrows & presence of critical erosion areas. -- OTHER 0	1,3,5,7,8, 9,10,11, 15,16,17, 18,19,20, 21,22,23, 24,25,26, 27,29,30, 31,32,33, 37,38,40, 45,46,54, 61,62,65, 69,70,73, 75,79,85, 87,95,97, 99,102
2. Runoff Potential	Low: -- Very flat to flat terrain (0-0.5% slope). -- Runoff curve number (RCN) 61 - 70. -- Dry, low rainfall (less than 20"). -- Even, gentle impact (scattered shower-type) rainfall. -- OTHER 10	Moderate: -- Flat to gently sloping (0.5-2.0% slope). -- RCN 71 - 80. -- Semidry (20-30"). -- Even, gentle to moderate intensity rainfall. -- OTHER 8	Considerable: -- Gently to moderately sloping (2.0-5.0% slope). -- RCN 81 - 90. -- Semiwet (30-40"). -- Even to uneven intense rainfall. -- OTHER 4	High: -- Moderately sloping to steep terrain (greater than 5%). -- RCN greater than 90. -- Wet (more than 40"). -- Intense uneven rainfall, especially in seasons when soil is exposed. -- OTHER 0	6,9,88,95
3. Filtering effect or sedimentation potential of a vegetated buffer or water/sediment collecting basin	-- Intervening vegetation between cropland & watercourse greater than 200 ft. -- Type of intervening vegetation ungrazed woodland, brush, or herbaceous plants. -- Water & sediment control basins properly installed & maintained. -- OTHER 8	-- Intervening vegetation between cropland & watercourse 100 to 200 ft. -- Type of intervening vegetation grazed woodland, brush, or herbaceous plants or range. -- Water & sediment control basins properly installed, but poorly maintained. -- OTHER 6	-- Intervening vegetation between cropland & watercourse 50 to 100 ft. -- Type of intervening vegetation high density cropland. -- Water & sediment control basins poorly installed & poorly maintained. -- OTHER 4	-- Cropping from less than 50 ft. up to water's edge. -- Type of intervening vegetation low density cropland or bare soil. -- No water & sediment control basins. -- OTHER 2	5,18,25, 27,79,107

4.	Resource management systems (RMS's) on whole farm (combined value for all agricultural areas)	-- Excellent management. -- RMS's always used as needed. -- OTHER 9	-- Good management. -- Most (80%) of the needed RMS's installed. -- OTHER 7	-- Fair management. -- About 50% of the needed RMS's installed. -- Cropping confined to proper land class. -- OTHER 3	-- Poor management. -- Few, if any, needed RMS's installed. -- Cropping not confined to proper classes. -- OTHER 0	Practices same as Rating Item #1
5.	Potential for ground water contamination	Low: -- Soils rich to very rich in organic matter (greater than 3.0%). -- Slow to very slow percolation in light textured soils such as clays, silty or sandy clays, or silty clay loams. -- Perched water table present. -- In protected bedrock areas (50 ft. of soil & shale cap), well depth is 75-100 ft. -- In protected bedrock areas overlain with 50 ft. of sand or gravel, well depth is greater than 150 ft. -- In shallow bedrock areas (25-50 ft. soil & shale cap), well depth greater than 200 ft. -- In Karst areas, well depth is greater than 1,000 ft., if aquifer is "confined." -- OTHER 9	Moderate: -- Soils rich to moderate in organic matter (3.0 to 1.5%). -- Slow to moderate percolation in clay loams or silts. -- Perched water table present. -- In protected bedrock areas, well depth is 30-74 ft. -- In protected bedrock areas overlain with 50 ft. of sand or gravel, well depth is 100-149 ft. -- In shallow bedrock areas, well depth is 50-199 ft. -- In Karst areas, well depth is 500-999 ft. -- OTHER 6	Considerable: -- Soils moderate to low in organic matter (1.5 to 0.5%). -- Moderate to rapid percolation in silty loams, loams, or silts. -- In protected bedrock areas, well depth is 15-29 ft. -- In protected bedrock areas overlain with 50 ft. of sand or gravel, well depth is 50 - 99 ft. -- In shallow bedrock areas, well depth is 25-49 ft. -- In Karst areas, well depth is 100-499 ft. -- OTHER 4	High: -- Soils low to very low in organic matter (less than 0.5%). -- Rapid percolation in coarse textured loamy sands or sands. -- In protected bedrock areas, well depth is less than 15 ft. -- In protected bedrock areas overlain with 50 ft. of sand or gravel, well depth is less than 50 ft. -- In shallow bedrock areas, well depth is less than 25 ft. -- In Karst areas, well depth is less than 100 ft. -- OTHER 0	See animal waste, nutrients, pesticide, & salt "B" Field Sheets for practices

1. Add the circled Rating Item scores to get a total for the field sheet. TOTAL [   ]
2. Check the ranking for this site based on the total field score. Check "excellent" if the score totals at least 40. Check "good" if the score falls between 26 and 39, etc. Record your total score and rank (excellent, good, etc.) in the upper right-hand corner of the field sheet. If a Rating Item is "fair" or "poor," find the practices in the right-hand column to help remedy the conditions.

RANKING   Excellent (40-46) [   ]   Good (26-39) [   ]   Fair (10-25) [   ]   Poor (9 or less) [   ]

Animal Waste

FIELD SHEET 2A: ANIMAL WASTE
INDICATORS FOR RECEIVING WATERCOURSES AND WATER BODIES

Evaluator _____ County/State _____ Date _____
Water Body Evaluated _____ Water Body Location _____ Total Score/Rank _____

(Circle one number among the four choices in each row which BEST describes the conditions of the watercourse or water body being evaluated. If a condition has characteristics of two categories, you can "split" a score.)

Rating Item	Excellent	Good	Fair	Poor
1. Evidence of animal waste: visual and olfactory	-:- No manure in or near water body. -:- No odor. -:- OTHER 9	-:- Occasional manure droppings where cattle cross or are in water. -:- Slight musk odor. -:- OTHER 6	-:- Manure droppings in concentrated localized areas. -:- Strong manure or ammonia odor. -:- OTHER 2	-:- Dry and wet manure all over banks or in water. -:- Strong manure & ammonia odor. -:- OTHER 0
2. Turbidity & color (observe in slow water)	-:- Clear or slightly greenish water in pond or along the whole reach of stream. -:- No noticeable colored film on submerged objects or rocks. -:- OTHER 9	-:- Occasionally turbid or cloudy. Water stirred up & muddy & brownish at animal crossings. -:- Pond water greenish. -:- Rocks or submerged objects covered with thin coating of green, olive, or brown build-up less than 5 mm thick. -:- OTHER 6	-:- Stream & pond water bubbly, brownish and cloudy where muddied by animal use. -:- Pea green color in ponds when not stirred up by animals. -:- Bottom covered w/ green or olive film. Rocks or submerged objects coated with heavy or filamentous build-up 5-75 mm thick or long. -:- OTHER 3	-:- Stream & pond water brown to black, occasionally with a manure crust along banks. -:- Sluggish & standing water—murky and bubbly (foaming). -:- Ponds often bright green or with brown/black decaying algal mats. -:- OTHER 0
3. Amount of aquatic vegetation	-:- Little vegetation—uncluttered look to stream or pond. -:- What you would expect for a pristine water body in area. -:- Usually fairly low amounts of many different kinds of plants. -:- OTHER 8	-:- Moderate amounts of vegetation; or -:- What you would expect for the naturally occurring site-specific conditions. -:- OTHER 6	-:- Cluttered weedy conditions. Vegetation sometimes luxurious and green. -:- Seasonal algal blooms. -:- OTHER 3	-:- Choked weedy conditions or heavy algal blooms or no vegetation at all. -:- Dense masses of slimy white, greyish green, rusty brown or black water molds common on bottom. -:- OTHER 0

4. Fish behavior in hot weather; fish kills, especially before dawn	:-- No fish piping or aberrant behavior. :-- No fish kills. :-- OTHER 8	:-- In hot climates, occasional fish piping or gulping for air in ponds just before dawn. :-- No fish kills in last two years. :-- OTHER 5	:-- Fish piping common just before dawn. :-- Occasional fish kills. :-- OTHER 3	:-- Pronounced fish piping. :-- Pond fish kills common. :-- Frequent stream fish kills during spring thaw. :-- Very tolerant species (e.g. bullhead, catfish). :-- OTHER 0
5. Bottom dwelling aquatic organisms	:-- Intolerant species occur: mayflies, stoneflies, caddisflies, water penny, riffle beetle and a mix of tolerants. :-- High diversity. :-- OTHER 9	:-- A mix of tolerants: shrimp, damselflies, dragonflies, black flies. :-- Intolerants rare. :-- Moderate diversity. :-- OTHER 5	:-- Many tolerants (snails, shrimp, damselflies, dragonflies, black flies). :-- Mainly tolerants and some very tolerants. :-- Intolerants rare. :-- Reduced diversity with occasional upsurges of tolerants, e.g. tube worms, and chironomids. :-- OTHER 3	:-- Only tolerants or very tolerants: midges, craneflies, horseflies, rat-tailed maggots, or none at all. :-- Very reduced diversity. upsurges of very tolerants common. :-- OTHER 0

1. Add the circled Rating Item scores to get a total for the field sheet.   TOTAL [   ]
2. Check the ranking for this site based on the total field score. Check "excellent" if the score totals at least 35. Check "good" if the score falls between 21 and 34, etc. Record your total score and rank (excellent, good, etc.) in the upper right hand corner of the field sheet. If a Rating Item is "fair" or "poor," complete Field Sheet $2B_1$ or $2B_2$.

RANKING[1]   Excellent (35-43) [   ]   Good (21-34) [   ]   Fair (7-20) [   ]   Poor (6 or less) [   ]

Animal Waste

FIELD SHEET 2B₁: ANIMAL WASTE
INDICATORS FOR PASTURE OR RANGE ANIMALS

Evaluator _____ County/State _____ Date _____

Field Evaluated _____ Field Location _____ Total Score/Rank _____

(Circle one number among the four choices in each row which BEST describes the conditions of the field or area being evaluated. If a condition has characteristics of two categories, you can "split" a score.)

Rating Item	Excellent	Good	Fair	Poor	Practices from Appendix E
1. Runoff Potential	Low: -- Runoff Curve Number (RCN) 61-70. -- Very flat to flat terrain (0-5% slope). -- Dry, low rainfall (less than 20") with rainfall erosivity (R) factor less than 50. -- Even, gentle impact (scattered shower-type) of rainfall. -- OTHER 10	Moderate: -- RCN 71-80. -- Flat to gently sloping (0.5-2.0% slope). -- Semidry (20-30") with R 50 to 100. -- Even, gentle to moderate intensity rainfall. -- OTHER 8	Considerable: -- RCN 81-90. -- Gently to moderately sloping (2-5% slope). -- Semiwet (30-40") with R 100 to 200. -- Even but intense rainfall. -- OTHER 4	High: -- RCN greater than 90. -- Moderately sloping to steep (greater than 5%). -- Wet (more than 40" rain) with R greater than 200. -- Intense uneven rainfall in seasons when soil is exposed. -- OTHER 0	14,16,17 18,19,20, 25,27,29, 33,38,40, 91,92,93 95,97
2. Ungrazed Buffer Zone	-- Pasture or range with a strip of intervening vegetation greater than 200 ft. -- OTHER 9	-- Pasture or range with 50 to 200 ft. strip of intervening vegetation. -- OTHER 7	-- Pasture or range with 10 to 50 ft. of intervening vegetation. -- OTHER 3	-- Pasture or range in close proximity to edge or adjacent to water course. -- OTHER 2	5,25,27
3. Rate of Waste Decomposition	-- Rapid decomposition of waste due to hot, sunny climate. -- OTHER 9	-- Moderate to rapid decomposition due to warm sunny climate. -- OTHER 7	-- Slow to moderate decomposition due to cooler, more overcast climate. -- OTHER 3	-- Slow decomposition due to cold climate with little direct solar radiation. -- OTHER 2	

		Excellent:	Good:	Fair:	Poor:	
4.	Pasture or Range Management	-- 90% cover. -- Proper grazing. -- Animal numbers within the carrying capacity of vegetation. -- No fertilization or pH adjustment and application of recommended amounts of fertilizer for maximum forage utilization based on soil tests. -- OTHER 9	-- 70-90% cover. -- Occasional bare areas. -- Animals exceed carrying capacity only 1 to 2 times per year. -- No fertilization or recommended amounts for maximum forage utilization. -- OTHER 6	-- 50-70% cover. -- Some bare spots. -- Animals exceed carrying capacity over 25% of the year. -- Fertilization at greater than recommended amounts for forage utilization. -- OTHER 3	-- 50% or less cover. -- Numerous bare spots. -- Animal numbers exceed carrying capacity 100% of year. -- Significant over-application of animal waste or commercial fertilizer close to water's edge. -- OTHER 0	5,15,18, 19,20,24, 25,27,29, 30,31,32, 33,38,40, 52,54,61, 65,69,70, 75,76,78, 79,92,95, 99,102, 105,106 107,108
		Low:	Moderate:	Considerable:	High:	
5.	Potential for ground water contamination	-- Soils rich to very rich in organic matter (greater than 3%). -- Slow to very slow percolation in light textured soils such as clays, silty or sandy clays, or silty clay loams. -- Perched water table present. -- In protected bedrock areas (50 ft. of soil & shale cap), well depth is 75-100 ft. -- In protected bedrock areas overlain with 50 ft. of sand or gravel, well depth is greater than 150 ft. -- In shallow bedrock areas (25-50 ft. soil & shale cap), well depth greater than 200 ft. -- In Karst areas, well depth is greater than 1,000 ft., if aquifer is "confined." -- OTHER 9	-- Soils rich to moderate in organic matter (1.5 to 3%). -- Slow to moderate percolation in clay loams or silts. -- Perched water table present. -- In protected bedrock areas, well depth is 30-74 ft. -- In protected bedrock areas overlain with 50 ft. of sand or gravel, well depth is 100-149 ft. -- In shallow bedrock areas, well depth is 50-199 ft. -- In Karst areas, well depth is 500-999 ft. -- OTHER 6	-- Soils moderate to low in organic matter (1.5 to 0.5%). -- Moderate to rapid percolation in silty loams, loams, or silts. -- In protected bedrock areas, well depth is 15-29 ft. -- In protected bedrock areas overlain with 50 ft. of sand or gravel, well depth is 50-99 ft. -- In shallow bedrock areas, well depth is 25-49 ft. -- In Karst areas, well depth is 100-499 ft. -- OTHER 4	-- Soils low to very low in organic matter (less than 0.5%). -- Rapid percolation in coarse textured loamy sands or sands. -- In protected bedrock areas, well depth is less than 15 ft. -- In protected bedrock areas overlain with gravel, well depth is less than 50 ft. -- In shallow bedrock areas, well depth is less than 25 ft. -- In Karst areas, well depth is less than 100 ft. -- OTHER 0	14,16,19, 25,27,30, 31,32,38, 40,45,54, 58,61,65, 97,102, 103,104, 105,106

TOTAL [   ]

1. Add the circled Rating Item scores to get a total for the field sheet.
2. Check the ranking for this site based on the total field score. Check "excellent" if the score totals at least 40. Check "good" if the score falls between 25 and 39, etc. Record your total score and rank (excellent, good, etc.) in the upper right-hand corner of the field sheet. If a Rating Item is "fair" or "poor," find the practices in the right-hand column to help remedy the conditions.

RANKING   Excellent (40-46) [   ]   Good (25-39) [   ]   Fair (10-24) [   ]   Poor (9 or less) [   ]

Animal Waste Page 1 of 2

FIELD SHEET 2B$_2$: ANIMAL WASTE
INDICATORS FOR TOTALLY OR PARTIALLY CONFINED ANIMALS

Evaluator _____ Field Location _____ Date _____
Field Evaluated _____ County/State _____ Total Score/Rank _____

(Circle one number among the four choices in each row which BEST describes the conditions of the field or area being evaluated. If a condition has characteristics of two categories, you can "split" a score.)

Rating Item	Excellent	Good	Fair	Poor	Practices from Appendix E
1. Runoff Potential	Low: -- Runoff Curve Number (RCN) 61-70. -- Very flat to flat terrain (0-0.5% slope). -- Dry, low rainfall (less than 20") with rainfall erosivity (R) factor less than 50. -- Even, gentle impact (scattered shower-type) of rainfall. -- OTHER 10	Moderate: -- RCN 71-80. -- Flat to gently sloping (0.5-2.0% slope). -- Semidry (20-30") with R 50 to 100. -- Even, gentle to moderate intensity rainfall. -- OTHER 8	Considerable: -- RCN 81-90. -- Gently to moderately sloping (2-5% slope). -- Semiwet (30-40") with R 100 to 200. -- Even but intense rainfall. -- OTHER 4	High: -- RCN greater than 90. -- Moderately sloping to steep (greater than 5%). -- Wet (more than 40" rain) with R greater than 200. -- Intense uneven rainfall in seasons when soil is exposed. -- OTHER 0	25,27,38, 69,70,78
2. Animal waste yield to water body; proportion of waste to leave the site	-- Site is 600 ft. from water body with intervening vegetation. -- Rapid decomposition of waste due to hot, sunny climate or low pH soils. -- OTHER 10	-- Site is between 200-500 ft. from water with intervening vegetation. -- Moderate to rapid decomposition due to warm, sunny climate. -- OTHER 8	-- Site 200 ft. from water. -- Slow to moderate decomposition due to cooler, more overcast climate. -- OTHER 4	-- Site is on bank of water body; or in close proximity to it. -- Slow decomposition due to cold climate with little direct solar radiation or high pH soils. -- OTHER 0	24,25,27, 38,40, 102,103, 104,105, 106
3. Animal access to water	-- None to very little. Watering areas located far from naturally occurring water bodies. -- OTHER 9	-- Very limited. Watering away from stream or pond. Stream used only as access path. -- OTHER 7	-- Access limited to watering. -- OTHER 3	-- Unlimited access for both watering and cooling. -- OTHER 0	1,40,54, 61,102

		Excellent management:		Good management:		Fair management:		Poor management	
4.	Runoff Management	Runoff is completely diverted away from concentrated waste. BMPs used as needed, such as surface water diversions, including guttering.	--	A good portion of clean runoff is diverted from waste. Runoff from feedlot, barns, etc. is diverted to holding pond.	--	Only a partial runoff management system. Evidence of contaminated runoff going directly to streams or ponds.	--	Little or no runoff management. Natural runoff removes most of the waste or little to no mgmt. of lagoons results in recurrent overflows. Evidence of lagoon overflows, manure-caked flow paths, etc.	25,27,38, 69,70,76, 78,102, 103,104, 105,106
		-- OTHER		-- OTHER		-- OTHER		-- OTHER	
		10		7		3		0	

Animal Waste Page 2 of 2

FIELD SHEET 2B₂: ANIMAL WASTE
INDICATORS FOR TOTALLY OR PARTIALLY CONFINED ANIMALS

5. Waste handling and utilization practices	Excellent mgmt. always with: -- Established collection schedule. -- Application at proper rates & times. -- Control of odor & pests. -- Regular sampling & record keeping. -- More than sufficient acreage for waste utilization. -- OTHER **10**	Good management most of the time (80%) with some of the following: -- Established collection schedules. -- Application at proper rates and times. -- Control of odor and pests. -- Sufficient acreage for waste utilization. -- OTHER **8**	Haphazard management common: -- Collection random. -- Applies waste anytime even before predicted rainfall. -- Odor and pests as occasional problems. -- Insufficient acreage for waste utilization. -- OTHER **4**	No or little management: -- A real mess most of the time. -- Continual odor and waste accumulation problems. -- OTHER **0**	102,103, 104,105, 106
6. Potential for ground water contamination	Low: -- Soils rich to very rich in organic matter (>3.0%). -- Slow to very slow percolation in light textured soils such as clays, silty or sandy clays, or silty clay loams. -- Perched water table present. -- In protected bedrock areas. (50 ft. of soil & shale cap), well depth is 75-100 ft. -- In protected bedrock areas overlain with 50 ft. of sand or gravel, well depth is greater than 150 ft. -- In shallow bedrock areas (25-50 ft. soil & shale cap), well depth greater than 200 ft. -- In Karst areas, well depth is greater than 1,000 ft., if aquifer is "confined." -- OTHER **9**	Moderate: -- Soils rich to moderate in organic matter (3.0 to 1.5%). -- Slow to moderate percolation in clay loams or silts. -- Perched water table present. -- In protected bedrock areas, well depth is 30-74 ft. -- In protected bedrock areas overlain with 50 ft. of sand or gravel, well depth is 100-149 ft. -- In shallow bedrock areas, well depth is 50-199 ft. -- In Karst areas, well depth is 500-999 ft. -- OTHER **6**	Considerable: -- Soils moderate to low in organic matter (1.5 to 0.5%). -- Moderate to rapid percolation in silty loams, loams, or silts. -- In protected bedrock areas, well depth is 15-29 ft. -- In protected bedrock areas overlain with 50 ft. of sand or gravel, well depth is 50-99 ft. -- In shallow bedrock areas, well depth is 25-49 ft. -- In Karst areas, well depth is 100-499 ft. -- OTHER **4**	High: -- Soils low to very low in organic matter (less than 0.5%). -- Rapid percolation in coarse textured loamy sands or sands. -- In protected bedrock areas, well depth is less than 15 ft. -- In protected bedrock areas overlain with 50 ft. of sand or gravel, well depth is less than 50 ft. -- In shallow bedrock areas, well depth is less than 25 ft. -- In Karst areas, well depth is less than 100 ft. -- OTHER **0**	14,16,19, 25,27,30, 31,32,38, 40,45,54, 58,61,65, 97,102, 103,104, 105,106

1. Add the circled Rating Item scores to get a total for the field sheet.
2. Check the ranking for this site based on the total field score. Check "excellent" if the score totals at least 51. Check "good" if the score falls between 33 and 50, etc. Record your total score and rank (excellent, good, etc.) in the upper right-hand corner of the field sheet. If a Rating Item is "fair" or "poor," find the practices in the right-hand column to help remedy the conditions.   TOTAL [   ]

RANKING   Excellent (51-58) [   ]   Good (33-50) [   ]   Fair (11-32) [   ]   Poor (10 or less) [   ]

Nutrients

FIELD SHEET 3A: NUTRIENTS
INDICATORS FOR RECEIVING WATERCOURSES AND WATER BODIES*

Evaluator _____ County/State _____ Date _____

Water Body Evaluated _____ Water Body Location _____ Total Score/Rank _____

(Circle one number among the four choices in each row which BEST describes the conditions of the watercourse or water body being evaluated. If a condition has characteristics of two categories, you can "split" a score.)

Rating Item	Excellent	Good	Fair	Poor
1. Total amount of aquatic vegetation at low flow or in pooled areas. Includes rooted and floating plants, algae, mosses & periphyton	-- Little vegetation, uncluttered look to stream or pond. OR What's expected for good water quality conditions in your region. -- Usually fairly low amounts of many different kinds of plants. -- OTHER **10**	-- Moderate amounts of vegetation. OR What's expected for good water quality conditions in your region. -- OTHER **6**	-- Cluttered weedy conditions. Vegetation sometimes luxurious and green. -- Seasonal algal blooms. -- OTHER **3**	-- Choked weedy conditions or heavy algal blooms or no vegetation at all. -- Dense masses of slimy white, greyish green, rusty brown or black water molds common on bottom. -- OTHER **0**
2. Color of water due to plants at base or low flow	-- Clear or slightly greenish water in pond or along the whole reach of stream. -- OTHER **9**	-- Fairly clear, slightly greenish. -- OTHER **6**	-- Greenish. Difficult to get pond sample without pieces of algae or weeds in it. -- OTHER **3**	-- Very, very green pond scums. -- Pea green color or pea soup condition during seasonal blooms of microscopic algae in ponds. -- "Oily-like" sheen when pea soup algae die off. -- OTHER **0**
3. Fish behavior in hot weather fish kills, especially before dawn	-- No fish piping or aberrant behavior. -- No fish kills. -- OTHER **9**	-- In hot climates, occasional fish piping or gulping for air in ponds just before dawn. -- No fish kills in last two years. -- OTHER **5**	-- Fish piping common just before dawn. -- Occasional fish kills. -- OTHER **3**	-- Pronounced fish piping. -- Pond fish kills common. -- Frequent stream fish kills during spring thaw. -- Very tolerant species (e.g. bullhead, catfish). -- OTHER **0**

4. Water use impacts; health effects for whole sub-watershed	-- None. -- OTHER 8	-- Minimal, such as reduced quality of fishing. -- OTHER 7	-- A couple of the following: -- Algal clogged pipes. -- Algal related taste, color, or odor problems with human or livestock water supply. -- Cattle abortion. -- Reduced recreational use due to weedy conditions, decay, odors, etc. -- OTHER 4	-- Several of the following: -- Algal clogged pipes. -- Algal related taste, color, or odor problems with human or livestock water supply. -- Cattle abortion. -- Reduced quality of fishery. -- Reduced recreational use due to weedy conditions, decay, odors, etc. -- Blue babies—incidence of methemoglobinemia due to high nitrate levels. -- Property devaluation. -- OTHER 2
5. Bottom-dwelling aquatic organisms	-- Intolerant species occur: mayflies, stoneflies, caddisflies, water penny, riffle beetle. -- High diversity. -- OTHER 9	-- Intolerants common. -- A mix of tolerants: shrimp, damselflies, dragonflies, black flies. -- Moderate diversity. -- OTHER 7	-- Mainly tolerants: snails, shrimp, damselflies, dragonflies, black flies. -- Mainly tolerants, but some very-tolerants. -- Intolerants rare. -- Reduced diversity with occasional upsurges of tolerants, e.g. tube worms, and chironomids. -- OTHER 3	-- Mainly very-tolerants: midges, craneflies, horseflies, rat-tailed maggots, or no organisms at all. -- Very reduced diversity, upsurges of very-tolerants common. -- OTHER 1

*The effects of nutrients may be "masked" by high sediment loads, creating sufficient turbidity to shade light-dependent aquatic vegetation. This may cause aquatic vegetation, a water quality indicator, to die and disappear from the watercourse. To obtain accurate nutrient levels in high sediment situations, chemical testing may be necessary. Under these circumstances you should contact a local or other water quality specialist.

1. Add the circled Rating Item scores to get a total for the field sheet. TOTAL [   ]
2. Check the ranking for this site based on the total field score. Check "excellent" if the score totals at least 38. Check "good" if the score falls between 23 and 37, etc. Record your total score and rank (excellent, good, etc.) in the upper right-hand corner of the field sheet. If a Rating Item is "fair" or "poor," complete Field Sheet 3B.

RANKING   Excellent (38-45) [   ]   Good (23-37) [   ]   Fair (9-22) [   ]   Poor (8 or less) [   ]

Nutrients Page 1 of 2

FIELD SHEET 3B: NUTRIENTS
INDICATORS FOR CROPLAND, HAYLAND, OR PASTURE

Evaluator _____   Field Location _____   County/State _____   Date _____
Field Evaluated _____   Total Score/Rank _____

(Circle one number among the four choices in each row which BEST describes the conditions of the field or area being evaluated. If a condition has characteristics of two categories, you can "split" a score.)

Rating Item	Excellent	Good	Fair	Poor	Practices from Appendix E
1. Erosion Potential	-- Not significant. -- Less than T (tolerance) little sheet & rill erosion, no gullies. -- Blocky, platy or massive soil structure. -- OTHER **10**	-- Some erosion evident. -- About T, some sheet & rill erosion. Very few gullies. -- Coarse granular to medium granular soils. -- OTHER **6**	-- Moderate erosion. -- T to 2T, gullies from heavy storm events obvious. -- Fine granular soils. -- Potentially highly erodible soils. -- OTHER **3**	-- Heavy erosion. -- More than 2T, many gullies and critical erosion areas. -- Very fine granular soils, highly erodible. -- OTHER **0**	7,9,10, 14,15,16, 17,18,20, 25,27,29, 30,45,61, 77,79,85, 87,97
2. Runoff Potential	Low: -- Soils hydraulic Group A. -- Very flat to flat terrain (0-0.5% slope). -- Dry, low rainfall (less than 20") with rainfall erosivity (R) factor less than 50. -- Even, gentle impact (scattered shower-type) of rainfall. -- OTHER **10**	Moderate: -- Soils Group B. -- Flat to gently sloping (0.5-2% slope). -- Semidry (20-30") with R 50 to 100. -- Even, gentle to moderate intensity rainfall. -- OTHER **8**	Considerable: -- Soils Group C. -- Gently to moderately sloping (2-5% slope). -- Semiwet (30-40") with R 100 to 200. -- Even but intense rainfall. -- OTHER **4**	High: -- Soils Group C. -- Moderately sloping to steep terrain (greater than 5%). -- Wet (more than 40" rain) with R greater than 200. -- Intense uneven rainfall in seasons when soil is exposed. -- OTHER **0**	6,9,52, 88,95
3. Resource Management Systems on whole farm (combined value for all agricultural areas—pastureland, cropland, or animal holding areas	-- Excellent management. -- RMSs always used as needed. -- OTHER **9**	-- Good management. -- Most (80%) of the needed RMSs installed. -- Predominance of farming practices diverting runoff away from receiving waters (terraces without tile drains). -- OTHER **7**	-- Fair management. -- About 50% of the needed RMSs installed. -- Cropping confined to proper land class. -- Predominance of farming practices diverting runoff toward receiving waters (tile drains and field ditches). -- OTHER **3**	-- Poor management. -- Few, if any, needed RMSs installed. -- Cropping not confined to proper classes. -- No diversion of runoff water; water flowing directly into receiving waters. -- OTHER **0**	All Practices

4. Buffer Zone	:-- Cropland is more than 600 ft. from water with intervening herbaceous vegetation (grass). :-- Cropland is more than 100 ft. from water with intervening woody vegetation (trees). :-- OTHER	:-- Cropland is less than 600 ft. but more than 200 ft. from water with intervening herbaceous vegetation (grass). :-- Cropland is less than 100 ft., but more than 50 ft. from water with intervening vegetation (trees). :-- OTHER	:-- Cropland is less than 200 ft. but more than 15 ft. from water with intervening herbaceous vegetation (grass). :-- Cropland is less than 50 ft., but more than 15 ft. from water with intervening woody vegetation (trees). :-- Little bank (riparian) vegetation. :-- OTHER	:-- Cropping up to the water's edge. :-- No bank (riparian) vegetation. :-- OTHER	25,26,27, 32,38
	10	7	2	0	

Nutrients Page 2 of 2

5. Fertilizer management practices	:-- Excellent management. :-- No fertilizer necessary. :-- Well defined schedule as to frequency & timing for inorganic or organic fertilizer depending on crop type, height of growth, etc. :-- Application of exactly the proper (recommended) amounts according to soil tests. Pays close attention to weather forecasts. Never applies before a storm. :-- Fertilizer is incorporated into the soil. :-- OTHER 9	:-- Good management. :-- Mainly follows a schedule but sometimes misses the best timing for the maximum utilization by the crop. :-- Usually follows directions for proper dosages of fertilizer and has soil tested regularly. Follows weather forecasts but once in a while will risk applying when rain is forecast. :-- Fertilizer is mainly of the incorporated slow-release type. :-- OTHER 7	:-- Haphazard management. :-- Follows a schedule about half the time. :-- Application is based on convenience. Tends to "overfertilize" by using more than the recommended dose as "insurance." :-- Occasionally loses much of application in a washout. :-- More than half the fertilizer is applied to the surface. :-- OTHER 3	:-- Little or erratic mgmt. :-- Seldom follows a schedule. :-- Applications without heed to weather forecasts. Often loses most of the applied fertilizer in a washout. Applies usually too little, sometimes too much. :-- Most of the fertilizer is surface applied without incorporation, e.g., in the North nitrogen application in the Autumn for some crops. :-- OTHER 0	7,8,9,12, 14,16,17, 25,27,30, 38,41,45, 52,60,69, 70,73,78, 79,80,81, 82,84,96

	Low:	Moderate:	Considerable:	High:	
6. Potential for ground water contamination	-- Soils rich to very rich in organic matter (> 3.0%). -- Slow to very slow percolation in light textured soils such as clays, silty or sandy clays, or silty clay loams. -- Perched water table present. -- In protected bedrock areas (50 ft. of soil & shale cap), well depth is 75-100 ft. -- In protected bedrock areas overlain with 50 ft. of sand or gravel, well depth is greater than 150 ft. -- In shallow bedrock areas (25-50 ft. soil & shale cap), well depth greater than 200 ft. -- In Karst areas, well depth is greater than 1,000 ft., if aquifer is "confined." -- OTHER	-- Soils rich to moderate in organic matter (3.0 to 1.5%). -- Slow to moderate percolation in clay loams or silts. -- Perched water table present. -- In protected bedrock areas, well depth is 30-74 ft. -- In protected bedrock areas overlain with 50 ft. of sand or gravel, well depth is 100-149 ft. -- In shallow bedrock areas, well depth is 50-199 ft. -- In Karst areas, well depth is 500-999 ft. -- OTHER	-- Soils moderate to low in organic matter (1.5 to 0.5%). -- Moderate to rapid percolation in silty loams, loams, or silts. -- In protected bedrock areas, well depth is 15-29 ft. -- In protected bedrock areas overlain with 50 ft. of sand or gravel, well depth is 50-99 ft. -- In shallow bedrock areas, well depth is 25-49 ft. -- In Karst areas, well depth is 100-499 ft. -- OTHER	-- Soils low to very low in organic matter (less than 0.5%). -- Rapid percolation in coarse textured loamy sand or sands. -- In protected bedrock areas, well depth is less than 15 ft. -- In protected bedrock areas overlain with 50 ft. of sand or gravel, well depth is less than 50 ft. -- In shallow bedrock areas, well depth is less than 25 ft. -- In Karst areas, well depth is less than 100 ft. -- OTHER	7,9,10,20, 25,27,30, 37,38,44, 60,64,80, 81,82,84, 87,96
	9	6	4	0	

1. Add the circled Rating Item scores to get a total for the field score.
2. Check the ranking for this site based on the total field score. Check "excellent" if the score totals at least 49. Check "good" if the score falls between 30 and 48, etc. Record your total score and rank (excellent, good, etc.) in the upper right-hand corner of the field sheet. If a Rating Item is "fair" or "poor," find the practices in the right-hand column to help remedy the conditions.

RANKING   Excellent (49-57) [   ]   Good (30-48) [   ]   Fair (9-29) [   ]   Poor (8 or less) [   ]   TOTAL [   ]

Pesticides

FIELD SHEET 4A: PESTICIDES
INDICATORS FOR RECEIVING WATERCOURSES AND WATER BODIES

Evaluator _____ County/State _____ Date _____

Water Body Evaluated _____ Water Body Location _____ Total Score/Rank _____

(Circle one number among the four choices in each row which BEST describes the conditions of the watercourse or water body being evaluated. If a condition has characteristics of two categories, you can "split" a score.)

Rating Item	Excellent	Good	Fair	Poor
1. Presence of pesticide containers	-- No containers in or near water. -- OTHER 9	-- No containers in or near water. -- OTHER 9	-- Containers located near the water. -- OTHER 5	-- Containers in the water. -- OTHER 3
2. Appearance of non-target vegetation	-- No leaf burn. -- No vegetation dieback. -- OTHER 9	-- Some leaf burn. -- No vegetation dieback. -- OTHER 6	-- Significant leaf burn. -- Some vegetation dieback. -- OTHER 4	-- Severe dieback of vegetation. -- OTHER 1
3. Overall diversity of insects, ("fish bait")	-- High diversity including dragonflies, stoneflies, mayflies, caddisflies, water mites or beetles. -- OTHER 10	-- Average diversity of insects—some of those listed under excellent. -- OTHER 8	-- Occasional insect kills. Reduced numbers and kinds. Upsurges of blackflies & chironomids. -- OTHER 3	-- Insect kills common. Not many fish-bait types such as hellgrammites (the larvae of dobsonflies), alderflies, or fishflies. -- OTHER 1
4. Overall diversity of fish	-- Excellent fish diversity—what's expected in the area. -- Presence of intolerants such as brook, brown or rainbow trout, salmon or stickleback. -- OTHER 9	-- Good fish diversity. -- Native salmonids (trout & salmon) begin to die out first. The least tolerant centrarchids (longear sunfish, rock bass, smallmouth bass, crappie, redfinned pickerel and bluegill) begin to decline. -- OTHER 7	-- Reduced fish diversity. -- The more tolerant centrarchids die off—blacknosed dace, common shiner, sculpin, creekchub, madtom, golden shiner, large mouth bass, blueback herring, and alewives. -- Larger proportion of green sunfish. -- Occasional (once every 1-2 years) pond fish kills. -- OTHER 4	-- Extremely reduced fish diversity. -- Only very tolerant species of cyprinids & ictalurids (such as brownhead carp, bullheads, white sucker, shad, and catfish). -- Some highly polluted waters (usually ponds) may lack fish entirely. -- OTHER 1

5. Fish kills; animal teratology (birth defects & tumors in fish & other animals)	:-- No fish kills in last 2 years. :-- No birth defects of tumors. :-- OTHER 9	:-- Fish kills rare in last 2 years. :-- Minimal birth defects & tumors occurring in the population randomly. :-- OTHER 5	:-- Occasional fish kills. :-- Some birth defects & tumors. :-- OTHER 3	:-- Fish kills common in last couple of years. :-- Frequent fish kills during spring thaws. :-- Yearly pond fish kills following aquatic vegetation dieback not uncommon. :-- Considerable numbers of birth defects & tumors. :-- OTHER 0
OPTIONAL 6. Fish behavior in hot weather; fish kills, especially before dawn	:-- Normal behavior, e.g. fish seen breaking the surface for insects. :-- No evidence of disease, tumors, fin damage or other anomalies. :-- No fish piping or aberrant behavior. :-- No fish kills. :-- OTHER 9	:-- Behavior as expected, e.g. evidence of fish, such as water rings or bubbles. :-- Little if any evidence of disease, tumors, fin damage, or other anomalies. :-- In hot climates, occasional fish piping or gulping for air in ponds just before dawn. :-- No fish kills in last 2 years. :-- OTHER 7	:-- Behavioral changes in fish—swimming near surface, uncoordinated movements, convulsive darting movements, erratic swimming up & down or in small circles, hyperexcitability (jumping out), difficulty in respiration. :-- More likely seen in ponds. :-- Fish piping common. :-- Occasional fish kills. :-- OTHER 4	:-- Fish avoidance or behaviors, such as erratic swimming near surface & gulping for or piping for air. More likely seen in ponds. :-- Pond fish kills common. :-- Frequent stream fish kills during Spring thaw. :-- Very tolerant species (e.g., bullhead, catfish). :-- OTHER 0

1. Add the circled Rating Item scores to get a total for the field score. TOTAL [ ]
2. Check the ranking for this site based on the total field score. Check "excellent" if the score totals at least 40. Check "good" if the score falls between 27 and 39, etc. Record your total score and rank (excellent, good, etc.) in the upper right-hand corner of the field sheet. If a Rating Item is "fair" or "poor," complete Field Sheet 4B.

RANKING   Excellent (40-46) [ ]   Good (27-39) [ ]   Fair (12-26) [ ]   Poor (11 or less) [ ]
OPTIONAL RANKING   Excellent (48-55) [ ]   Good (32-47) [ ]   Fair (14-31) [ ]   Poor (13 or less) [ ]

Pesticides Page 1 of 2

FIELD SHEET 4B: PESTICIDES
INDICATORS FOR CROPLAND, HAYLAND, OR PASTURE

Evaluator _____ Date _____
Field Evaluated _____ County/State _____ Total Score/Rank _____
Field Location _____

(Circle one number among the four choices in each row which BEST describes the conditions of the field or area being evaluated. If a condition has characteristics of two categories, you can "split" a score.)

Rating Item	Excellent	Good	Fair	Poor	Practices from Appendix E
1. Erosion Potential	-- Not significant. -- Less than T (tolerance), little sheet, rill, or furrow erosion. -- No gullies. -- OTHER 10	-- Some erosion evident. -- About T; some sheet, rill, or furrow erosion. -- Very few gullies. -- OTHER 7	-- Moderate erosion. -- T to 2T. -- Gullies or furrows from heavy storm events obvious. -- OTHER 3	-- Heavy erosion. -- Greater than 2T. -- Many gullies or furrows & presence of critical erosion areas. -- OTHER 0	7,9,10,16, 17,27,29, 39,45,46, 55,74,77, 79,87,95
2. Buffer Zone	-- Intervening vegetation between cropland & watercourse greater than 200 ft. -- Type of intervening vegetation ungrazed woodland, brush, or herbaceous plants. -- OTHER 8	-- Intervening vegetation between cropland & watercourse 100 to 200 ft. -- Type of intervening vegetation grazed woodland, brush, or herbaceous plants or range. -- OTHER 6	-- Intervening vegetation between cropland & watercourse 50 to 100 ft. -- Type of intervening vegetation high density cropland. -- OTHER 4	-- Cropping from less than 50 ft. up to water's edge. -- Type of intervening vegetation low density cropland or bare soil. -- OTHER 2	25,27,38
3. Appearance of non-target vegetation	-- No leaf burn. -- No evidence of dieback. -- OTHER 9	-- Some leaf burn. -- No dieback. -- OTHER 6	-- Significant leaf burn. -- Some vegetation dieback. -- OTHER 4	-- Severe dieback of vegetation. -- OTHER 1	2,4,34,41, 42,43,48, 49,50,51, 66,72
4. Runoff Potential	Low: -- Runoff Curve Number (RCN) 61-70. -- Very flat to flat terrain (0-0.5% slope). -- Dry, low rainfall (less than 20") with rainfall erosivity (R) factor less than 50. -- Even, gentle impact (scattered shower-type) of rainfall. -- OTHER 10	Moderate: -- RCN 71-80. -- Flat to gently sloping (0.5-2.0% slope). -- Semidry (20-30") with R 50 to 100. -- Even, gentle to moderate intensity rainfall. -- OTHER 8	Considerable: -- RCN 81-90. -- Gently to moderately sloping (2-5% slope). -- Semiwet (30-40") with R 100 to 200. -- Even but intense rainfall. -- OTHER 4	High: -- RCN greater than 90. -- Moderately sloping to steep (greater than 5%). -- Wet (more than 40" rain) with R greater than 200. -- Intense uneven rainfall in seasons when soil is exposed. -- OTHER 0	6,9,52, 88,95

| 5. Type of pesticide | :-- Narrow spectrum, species specific. :-- Water soluble, very rapidly degrading. :-- OTHER 8 | :-- Fairly narrow range of toxicity. :-- Water soluble, rapid-to-moderate degradation. :-- OTHER 5 | :-- Persistent, not species specific. :-- Fat soluble, nonbio-degradable. :-- OTHER 3 | :-- Persistent, wide spectrum biocide (harms "any living thing"). :-- Fat soluble, nonbiode-gradable. :-- OTHER 1 | 2,49,66 |

Pesticides

6. Pesticide management including amount of pesticide applied per acre; the frequency of application, timing & manner of application; and clean-up practices	-- Application according to a well defined pest management program such as integrated Pest Management (IPM) with close supervision by professional. -- Insecticides applied once every two years. One herbicide treatment per year. -- Careful nonaerial spraying or incorporating into the soil. -- Spraying on dry, hot, windless days. -- Follows instructions on pesticide label. Discards containers at appropriate disposal centers. -- Uses a professional applicator.	-- Application of recommended dosages by certified applicators based on scouting by professionals. -- Insecticides applied twice per year. Two herbicide treatments per year. OR Insecticides & herbicides applied as needed. -- Careful non-aerial or aerial spraying. -- Spraying on calm, dry days. -- Careful to avoid spills. Careful to keep containers away from water body.	-- Application based on scouting done by the landowner; extra pesticide beyond the recommended dosage to insure pest control. -- Insecticides applied 2 to 5 times per year. 2 to 3 herbicide treatments per year. -- Casual non-aerial or aerial spraying. -- Spraying with minimal concern about weather. -- Containers discarded haphazardly. Containers washed in a water body or in close proximity to the water, so that contamination is likely.	-- Application by a schedule that meets the needs of the landowner. No scouting. -- Landowner strives for zero pests (complete eradication) by doubling or more than doubling the application rate. -- Insecticides applied more than 5 times per year. More than 3 herbicide treatments per year. -- Application almost exclusively aerial. -- Spraying with no heed to the weather. Application on windy, rainy, days common. -- Careless discarding of containers in water bodies or sinkholes. Doesn't heed warnings for human safety with regard to application, cleanup, and disposal.	2,4,9,10, 13,16,17, 20,21,25, 27,29,33, 34,38,41, 42,43,45, 47,48,49, 50,51,55, 59,66,72, 77,80,87
	-- OTHER 10	-- OTHER 7	-- OTHER 3	-- OTHER 0	

7. Potential for ground water contamination	Low:	Moderate:	Considerable:	High:	14,16,19, 25,27,30, 31,32,38, 40,45,54, 58,61,65, 97,102, 103,104, 105,106
	-- Soils rich to very rich in organic matter (>3.0%).	-- Soils rich to moderate in organic matter (3.0 to 1.5%).	-- Soils moderate to low in organic matter (1.5 to 0.5%).	-- Soils low to very low in organic matter (less than 0.5%).	
	-- Slow to very slow percolation in light textured soils such as clays, silty or sandy clays, or silty clay loams.	-- Slow to moderate percolation in clay loams or silts.	-- Moderate to rapid percolation in silty loams, loams, or silts.	-- Rapid percolation in coarse textured loamy sands or sands.	
	-- Perched water table present.	-- Perched water table present.			
	-- In protected bedrock areas (50 ft. of soil & shale cap), well depth is 75-100 ft.	-- In protected bedrock areas, well depth is 30-74 ft.	-- In protected bedrock areas, well depth is 15-29 ft.	-- In protected bedrock areas, well depth is less than 15 ft.	
	-- In protected bedrock areas overlain with 50 ft. of sand or gravel, well depth is greater than 150 ft.	-- In protected bedrock areas overlain with 50 ft. of sand or gravel, well depth is 100-149 ft.	-- In protected bedrock areas overlain with 50 ft. of sand or gravel, well depth is 50-99 ft.	-- In protected bedrock areas overlain with 50 ft. of sand or gravel, well depth is less than 50 ft.	
	-- In shallow bedrock areas (25-50 ft. soil & shale cap), well depth greater than 200 ft.	-- In shallow bedrock areas, well depth is 100-199 ft.	-- In shallow bedrock areas, well depth is 25-49 ft.	-- In shallow bedrock areas, well depth is less than 25 ft.	
	-- In Karst areas, well depth is greater than 1,000 ft., if aquifer is "confined."	-- In Karst areas, well depth is 500-999 ft.	-- In Karst areas, well depth is 100-499 ft.	-- In Karst areas, well depth is less than 100 ft.	
	-- OTHER	-- OTHER	-- OTHER	-- OTHER	
	9	6	4	0	

1. Add the circled Rating Item scores to get a total for the field sheet.
2. Check the ranking for this site based on the total field score. Check "excellent" if the score totals at least 54. Check "good" if the score falls between 35 and 53, etc. Record your total score and rank (excellent, good, etc.) in the upper right-hand corner of the field sheet. If a Rating Item is "fair" or "poor," find the practices in the right-hand column to help remedy the conditions.

RANKING Excellent (54-64) [ ]   Good (35-53) [ ]   Fair (14-34) [ ]   Poor (13 or less) [ ]   TOTAL [ ]

Salinity

FIELD SHEET 5A: SALINITY
INDICATORS FOR RECEIVING WATERCOURSES AND WATER BODIES

Evaluator _____ County/State _____ Date _____

Water Body Evaluated _____ Water Body Location _____ Total Score/Rank _____

(Circle one number among the four choices in each row which BEST describes the conditions of the watercourse or water body being evaluated. If a condition has characteristics of two categories, you can "split" a score.)

Rating Item	Excellent	Good	Fair	Poor
1. Geology of area and geochemistry of water	-- Agricultural area overlies formations of igneous or metamorphic origin. -- Few fractures or faults in the area. -- Very low to low mineral content—soft waters of the East and Southeast. -- OTHER 10	-- Agricultural area primarily overlies formations of igneous or metamorphic origin with occasional areas above marine deposits. -- Few fractures or faults. -- Low to moderate mineral content—soft waters. -- OTHER 7	-- Agricultural area overlies marine deposits. Faulting common. -- Moderate to high mineral content—hard waters of mountain states, deserts, and Great Plains. -- OTHER 3	-- Agricultural area overlies marine deposits of recent origin. -- Fractures and faulting very common in the area. -- High to very high mineral content. Soils of marine origin. Salty ground water & springs. Mineral springs. -- OTHER 0
2. Precipitation and irrigation requirements	-- Average crop water consumption is equal to or less than average precipitation. -- Minimal irrigation required. -- OTHER 8	-- Average crop water consumption is between 5 & 10% more than average precipitation. -- Moderate irrigation req'd. -- OTHER 6	-- Average crop water consumption is between 10 & 25% more than precipitation. -- Considerable irrigation required. -- OTHER 4	-- Average crop water consumption exceeds average precipitation by more than 25%. -- Maximal irrigation required. -- OTHER 0
3. Location of watercourse in watershed	-- Near headwaters. -- OTHER 9	-- Not far from headwaters. -- OTHER 7	-- Moderate distance from headwaters. -- OTHER 5	-- Far from headwaters. -- OTHER 3
4. Appearance of water's edge (shoreline or banks)	-- No evidence of salt crusts. -- OTHER 9	-- Some evidence of white, crusty deposits on banks. -- OTHER 6	-- Numerous localized patches of white, crusty deposits on banks. -- OTHER 4	-- Most of the pond or stream bank covered with a thick, white, crusty deposit. Salt "feathering" on posts abundant. -- OTHER 1

5. Appearance of aquatic vegetation	:-- No evidence of wilting, toxicity, or stunting.	:-- Minimal wilting and toxicity, bleaching, leaf burn. Little if any stunting.	:-- Stream or pond vegetation may show wilted and toxic symptoms—bleaching, leaf burn. Presence of some salt-tolerant species.	:-- Evidence of severe wilting, toxicity, or stunting. Presence of only the most salt-tolerant species or complete absence of vegetation.
	:-- OTHER	:-- OTHER	:-- OTHER	:-- OTHER
	10	7	3	0
6. Streamside vegetation	:-- Very few species.	:-- Few salt tolerant species. Refer to list below*.	:-- Increasing dominance of salt-tolerant species.	:-- Vegetation almost totally salt-tolerant species or absence of vegetation.
	:-- OTHER	:-- OTHER	:-- OTHER	:-- OTHER
	8	7	5	3
OPTIONAL				
7. Animal teratology (birth defects & tumors in fish and other animals)	:-- No birth defects or tumors.	:-- Minimal birth defects & tumors occuring in the population randomly.	:-- Some birth defects & tumors.	:-- Considerable numbers of birth defects & tumors.
	:-- OTHER	:-- OTHER	:-- OTHER	:-- OTHER
	10	6	1	0

*Salt-tolerant species include greasewood, alkali sacaton, fourwing saltbush, shadscales, saltgrass, tamarisk (salt cedar), galleta, western wheatgrass, crested wheat, mat saltbush, reed canarygrass, and rabbitbrush.

1. Add the circled Rating Item scores to get a total for the field sheet. TOTAL [   ]
2. Check the ranking for this site based on the total field score. Check "excellent" if the score totals at least 47. Check "good" if the score falls between 32 and 46, etc. Record your total score and rank (excellent, good, etc. in the upper right-hand corner of the field sheet. If a Rating Item is "fair" or "poor", complete Field Sheet $5B_1$ or $5B_2$.

RANKING   Excellent (47-54) [   ]   Good (32-46) [   ]   Fair (15-31) [   ]   Poor (14 or less) [   ]
RANKING (OPTIONAL) Excellent (55-64) [   ]   Good (35-54) [   ]   Fair (16-34) [   ]   Poor (15 or less) [   ]

Salinity Indicators Page 1 of 2

FIELD SHEET 5B₁: SALINITY INDICATORS

Evaluator _____ Field Location _____ County/State _____ Date _____
Field Evaluated _____ Total Score/Rank _____

(Circle one number among the four choices in each row which BEST describes the conditions of the field or area being evaluated. If a condition has characteristics of two categories, you can "split" a score.)

Rating Item	Excellent	Good	Fair	Poor	Practices from Appendix E
1. Length of off-farm delivery system from headgate to farm boundary	-- Less than ¼ mile. -- OTHER 10	-- Between ¼ and ½ mile. -- OTHER 7	-- Between ½ and 1 mile. -- OTHER 3	-- Greater than 1 mile. -- OTHER 0	
2. Irrigation management practices, including: seepage potential of delivery system, overall irrigation rating, and timing of irrigation	-- All canals lined or piped. -- Excellent maintenance. -- Clay soil texture. -- Seepage rate of 0.1 to 1.0 cu. ft. of water per sq. ft. of surface per day ($ft^3/ft^2/day$). -- Sediment ponds, fertilizer management, monitoring flow, and other BMPs used as needed. -- Irrigation scheduling based on crop needs and testing by tensiometer, moisture block or neutron probe. -- OTHER 10	-- Canals are partially lined. -- Moderate maintenance. -- Sandy clay soil texture. -- Seepage rate of 0.2 to 1.1 $ft^3/ft^2/day$. -- Most (80%) of needed practices installed. -- Timing based on crop needs and maximum allowable deficiency (e.g., testing by wet ball or soil probe). -- OTHER 7	-- Vegetated canals. -- Little maintenance. -- Sandy, silty, clay loams. -- Seepage rate 0.3 to 1.3 $ft^3/ft^2/day$. -- About 50% of needed practices installed. -- Irrigation tied to traditional irrigation scheduling with little regard to crops' water requirements. -- OTHER 3	-- Earthen canals. -- Maintenance leading to disturbed canal bottom. -- Sands, loams, & silty loams. -- Seepage rate 0.5 to 1.5 $ft^3/ft^2/day$. -- Poor management. Few needed practices installed. Continuing increase in number of evaporation ponds. -- Excessive irrigation based on convenience & traditional irrigation scheduling. No consideration of crop needs. -- OTHER 0	21,35,36, 37,38,44, 53,59,62, 67,68,69, 70,71,73, 98,101, 108,109

3. Kind & properties of soils; permeability (adjusted Sodium Adsorption Ratio-SAR)	:-- Coarse textured particles. Deep topsoil—excellent tilth.	:-- No restrictive properties— good tilth.	:-- Clay soils with high sodium & high salt. Reduced tilth. Several of the characteristics listed under poor. :-- Montmorillonite clay with SAR = 8. :-- Illite clay with SAR of 12-15. :-- Kaolinite clay with SAR of 20-23.	:-- High montmorillonite clays with high sodium & high salt. Black soils with dissolved organic matter. Poor tilth. Puddling, soggy soils, poor infiltration and drainage. Slick spots and white crust. :-- Montmorillonite clay with SAR 9. :-- Illite clay with SAR 16. :-- Kaolinite clay with SAR 24.	
	:-- OTHER	:-- OTHER	:-- OTHER	:-- OTHER	68
	9	6	3	0	
4. Soil salinity (mmhos/cm) or (Decisiemans/ meter)	:-- Less than 0.8 (mmhos/cm).	:-- Between 0.8 & 1.5 (mmhos/cm).	:-- Between 1.5 & 2.5 (mmhos/cm).	:-- Greater than 2.5 (mmhos/cm).	
	:-- OTHER	:-- OTHER	:-- OTHER	:-- OTHER	
	9	6	3	0	

Salinity Indicators Page 2 of 2

5. Crop type. Productivity and appearance, including specific ion toxicity (varies with species sensitivity to particular toxin)	:-- Crop type relatively non-tolerant to salt. Refer to Appendix. :-- High productivity. :-- Prolific growth. :-- None. :-- OTHER 9	:-- Moderately salt-tolerant species predominate. :-- Average productivity—what's expected in the region. :-- Some wilting. :-- OTHER 6	:-- Less salt-tolerant crops die out. Replacement by relatively salt-tolerant species. :-- Less than expected productivity. Some stunting. :-- Wilted & noticeable toxic symptoms—tip and marginal leaf burn, chlorosis (bleached areas); defoliation. Deep blue-green foliage. Thickened waxy coating on leaves. :-- OTHER 3	:-- Only highly salt-tolerant crops can be grown. :-- Plants of variable size. Stunted growth. Reduced production. :-- Toxic symptoms & dieoff of crops sensitive to given ions. :-- OTHER 1	17,22
6. Animal productivity and health	:-- No reduction in productivity. :-- No incidence of disease. :-- OTHER 9	:-- Minimal reduction in productivity. :-- Minimal incidence of disease. :-- OTHER 6	:-- Some reduction in total growth, milk production, etc. :-- Moderate incidence of disease symptoms such as diarrhea. :-- OTHER 3	:-- Greatly reduced growth, milk production, etc. :-- With sudden salinity changes, livestock may reject water. :-- High incidence of disease symptoms such as diarrhea. :-- OTHER 1	24,40,54, 64,71,72, 83

	Low:	Moderate:	Considerable:	High:	
7. Potential for ground water contamination	-- Soils rich to very rich in organic matter (> 3.0%). -- Slow to very slow percolation in light textured soils such as clays, silty or sandy clays, or silty clay loams. -- Perched water table present. -- In protected bedrock areas (50 ft. of soil & shale cap), well depth is 75-100 ft. -- In protected bedrock areas overlain with 50 ft. of sand or gravel, well depth is greater than 150 ft. -- In shallow bedrock areas (25-50 ft. soil & shale cap), well depth greater than 200 ft. -- In Karst areas, well depth is greater than 1,000 ft., if aquifer is "confined." -- OTHER	-- Soils rich to moderate in organic matter (3.0 to 1.5%). -- Slow to moderate percolation in clay loams or silts. -- Perched water table present. -- In protected bedrock areas, well depth is 30-74 ft. -- In protected bedrock areas overlain with 50 ft. of sand or gravel, well depth is 100-149 ft. -- In shallow bedrock areas, well depth is 50-199 ft. -- In Karst areas, well depth is 500-999 ft. -- OTHER	-- Soils moderate to low in organic matter (1.5 to 0.5%). -- Moderate to rapid percolation in silty loams, loams, or silts. -- In protected bedrock areas, well depth is 15-29 ft. -- In protected bedrock areas overlain with 50 ft. of sand or gravel, well depth is 50-99 ft. -- In shallow bedrock areas, well depth is 25-49 ft. -- In Karst areas, well depth is 100-499 ft. -- OTHER	-- Soils low to very low in organic matter (less than 0.5%). -- Rapid percolation in coarse textured loamy sands or sands. -- In protected bedrock areas, well depth is less than 15 ft. -- In protected bedrock areas overlain with 50 ft. of sand or gravel, well depth is less than 50 ft. -- In shallow bedrock areas, well depth is less than 25 ft. -- In Karst areas, well depth is less than 100 ft. -- OTHER	21,35,36, 37,53,62, 64,68,70, 73,91,92
	9	6	4	0	

1. Add the circled Rating Item scores to get a total for the field sheet.
2. Check the ranking for this site based on the total field score. Check "excellent" if the score totals at least 54. Check "good" if the score falls between 31 and 53, etc. Record your total score and rank (excellent, good, etc.) in the upper right-hand corner of the field sheet. If a Rating Item is "fair" or "poor," find the practices in the right-hand column to help remedy the conditions.

TOTAL [ ]

RANKING   Excellent (54-65) [ ]   Good (33-53) [ ]   Fair (12-32) [ ]   Poor (11 or less) [ ]

Salinity

FIELD SHEET 5B₂: SALINITY
INDICATORS FOR SALINE SEEPS

Evaluator _____ : Seep Location _____ : County/State _____ : Date _____
Saline Seep Evaluated _____ : Total Score/Rank _____

(Circle one number among the four choices in each row which BEST describes the conditions of the field or area being evaluated. If a condition has characteristics of two categories, you can "split" a score.)

Rating Item	Excellent	Good	Fair	Poor	Practices from Appendix E
1. Geology	-- Agricultural area overlies formations of igneous or metamorphic origin. -- Few fractures or faults in the area. -- OTHER 10	-- Agricultural areas primarily overlies formations of igneous or metamorphic origin with occasional areas above marine deposits. -- Few fractures or faults. -- OTHER 7	-- Agricultural area overlies marine deposits. -- Faulting common. -- OTHER 3	-- Agricultural area overlies marine deposits of recent origin. -- Fractures and faulting very common in the area. -- OTHER 0	
2. Precipitation & irrigation requirements	-- Average crop water consumption is equal to or less than average precipitation. -- OTHER 8	-- Average crop water consumption is between 5 and 10% more than average precipitation. -- OTHER 6	-- Average crop water consumption is between 10 and 25% more than precipitation. -- OTHER 4	-- Average crop water consumption exceeds average precipitation by more than 25%. -- OTHER 0	
3. Cropping system	-- Crop rotation consists of annual crops with maximum consumptive water use. -- OTHER 8	-- Crop rotation consists of annual crops. -- OTHER 6	-- Crop rotation contains a biannual fallow period. -- Crops with maximum water consumptive use grown. -- OTHER 4	-- Crop rotation contains a biannual fallow period. -- OTHER 2	17,37,68, 72
4. Field appearance, including salt crusts	-- Downslope fields exhibit even-appearing crop growth. High yields are common for the area. -- OTHER 9	-- Downslope areas of field or downslope fields exhibit even crop growth, but of reduced yield for the area. -- OTHER 7	-- Downslope areas of field or downslope fields have uneven growth of crops with patches of crops significantly stunted. -- Occasionally white crust occurs in these patches. -- OTHER 3	-- Downslope areas of fields have bare spots not accounted for by soil variations. Bare spots occur in highly saline soils. White crust common. -- OTHER 1	

1. Add the circled Rating Item scores to get a total for the field sheet.	TOTAL [ ]
2. Check the ranking for this site based on the total field score. Check "excellent" if the score totals at least 30. Check "good" if the score falls between 20 and 29, etc. Record your total score and rank (excellent, good, etc.) in the upper right-hand corner of the field sheet. If a Rating Item is "fair" or "poor," find the practices in the right-hand column to help remedy the conditions.

RANKING   Excellent (30-35) [   ]   Good (20-29) [   ]   Fair (8-19) [   ]   Poor (7 or less) [   ]